FM 3-61.1

Field Manual
No. 3-61.1

Headquarters
Department of the Army
Washington, DC, 1 October 2000

Public Affairs Tactics, Techniques and Procedures

Contents

Preface

The mission of Army Public Affairs (PA) is to fulfill the Army's obligation to keep the American people and the Army informed, and to help establish the conditions that lead to confidence in America's Army and its readiness to conduct operations in peacetime, conflict and war. PA is a critical battlefield function in today's global information environment since every aspect of an Army operation is subject to instantaneous scrutiny.

This field manual (FM) sets forth tactics, techniques and procedures (TTPs) for conducting PA operations in accordance with the doctrinal principles contained in FM 3-0 (100-5), *Operations* and FM 3-61 (46-1), *Public Affairs Operations*. Although this manual is primarily designed to be used by public affairs officers, noncommissioned officers and civilians, it presents TTPs that all leaders conducting operations in the information age should be familiar with.

FM 3-61-1 (46-1-1) is applicable to units and individuals in both the active and reserve components. It serves as a foundation for integrating PA into Army doctrine, training, leader development, organization, materiel and soldier initiatives. In conjunction with the Army Training and Evaluation Program (ARTEP) and other training guidance, it should also be used to plan, integrate and execute individual and collective PA training in units throughout the Army.

The proponent for this manual is the Office of the Chief of Public Affairs. Send comments and recommendations on DA Form 2028 to Director, Army Public Affairs Center, ATTN: SAPA-PA, Fort Meade, MD 20755-5650.

Unless this publication states otherwise, masculine nouns or pronouns do not refer exclusively to men.

This document is approved for public release; distribution is unlimited.

Introduction

With the extremely sophisticated communication technologies of the global information environment (GIE), the nature of media coverage has a significant impact on the conduct of war and stability and support operations at the strategic, operational and tactical levels. Civilian and military news media coverage influences the perceptions of soldiers, family members, the public and political leaders, and affects the direction, range and duration of operations. It has a direct relation to the confidence these key audiences have in the Army and its execution of operations.

Effective PA operations are critical to successful Army operations in the information age. They assist the commander in monitoring and understanding public opinion, explaining the situational context of events and communicating the Army's perspective clearly and without filters. They enable the commander to interpret the perceptions of external and internal audiences and influence the way in which discussion of the operation is framed.

Synchronized, well-planned and actively executed PA tactics, techniques and procedures significantly clear the fog of war and impact the morale and effectiveness of the force. They reduce distractions, misinformation, uncertainty, confusion and other factors that cause stress and undermine efficient operations. They enhance understanding, acceptance and support. Effective PA operations contribute to soldier confidence, discipline, will to win, and unit cohesion.

FM 3-61-1 (46-1-1), *Public Affairs Tactics, Techniques and Procedures*, builds on the doctrinal foundation of FM 46-1. It translates the PA fundamentals and principles into detailed guidance for the planning, coordination and execution of PA operations. It provides what is required for the information age -- a sophisticated approach to conducting PA operations. It is the TTP that brings Army Public Affairs into the "information age."

FM 3-61 (46-1), *Public Affairs Operations*, addresses fundamental PA concepts in depth, and provides the linkage between PA and the Army's keystone doctrine, FM 3-0 (100-5), *Operations*. It recognizes that a refocused, restructured military will conduct operations in an information environment in which detailed, graphic, and live coverage of events are transmitted around the world. It builds from the understanding that information availability will influence strategic decisions and the direction, range and duration of operations.

FM 3-61 (46-1) also examines PA operations at the different levels of war and across the range of operations. It discusses PA operations with respect to the Principles of War and the Tenets of Army Operations. It analyzes the PA contributions to build and sustain combat power, defines the PA mission, and establishes strategic PA goals, fundamental PA principles, and underlying considerations for planning integrated information strategies.

Public affairs frequently deals in intangibles -- perceptions and implications -- that are not easily quantifiable or qualifiable, but are essential to commanders. The PA objectives, processes and methods presented in FM 3-61-1 (46-1-1) will assist Army leaders and PA professionals to develop solutions to the complex PA issues they will confront.

Chapter 1
Public Affairs Fundamentals

Public Affairs fulfills the Army's obligation to keep the American people and the Army informed and helps establish the conditions that lead to confidence in America's Army and its readiness to conduct operations in peacetime, conflict and war.

FM 3-61 (46-1), Public Affairs Operations

THE PA MISSION

1-1. The American public, internal Army audiences, allies, adversaries and other critical audiences have access to an ever expanding array of public and military media. Newspapers, magazines, radio, television and electronic media are independent conduits of information. They provide news, analysis, interpretation and commentary and serve as a forum for ideas, opinions and public debate. What appears in the media, both civilian and military, shapes perceptions, attitudes and opinions, and can have a direct impact on mission success.

1-2. The vast majority of both civilian and military media representatives are committed to providing responsible, accurate, balanced coverage. Although there are exceptions, most media representatives, even when editorializing, are focused on achieving a credible presentation. To accomplish this, media representatives investigate issues, ask tough, challenging questions, and pursue verifiable answers. They seek information, interpretation and perspective. Army leaders at all levels need to educate media representatives and facilitate their efforts to provide an accurate, balanced and credible presentation of timely information.

1-3. Army leaders do this by integrating public affairs into the planning process and synchronizing PA operations with other facets of the operation. Integrating and synchronizing public affairs issues allows commanders to communicate their perspective and achieve a balanced, accurate, credible information presentation.

1-4. The PA mission identifies the essential contribution that PA makes to America's Army. The mission and the strategic goals derived from it provide the foundation on which public affairs operations are built. Strategies, which are developed from the perspective that every aspect of every operation could become an issue of interest in the global information environment, are the most successful. Developing such strategies requires that PA personnel access, analyze and anticipate potential issues by conducting a thorough mission analysis.

1-5. The challenge for commanders, and personnel supporting them, is to plan and execute operations, which accomplish this mission and support these goals. To do this, PA must be integrated into the planning and decision-making process from receipt of the mission.

1-6. The need to integrate and synchronize PA early derives from the fact that in most situations media representatives will be present in an area of operations before the arrival of Army forces. They will know the area of operations and because they are covering the story as it evolves, will have an understanding of, and opinion about, the situation.

1-7. Media interest will normally be the most intense at the onset of operations. Media representatives will cover the deployment of Army forces, their arrival in the area and their initial conduct. To support the commander and the force in their interactions with media representatives during these early stages, public affairs personnel should be deployed in the first days, if not hours, of the operation. Dealing with a large international press corps constitutes the most immediate public affairs challenge facing the commander during contingency.

PUBLIC AFFAIRS FORCE

1-8. The changing information environment in which the Army conducts war and stability and support operations makes it necessary for PA officers, NCOs and specialists to respond to increasingly complex,demanding challenges. They must be prepared to support the commander with a wide range of knowledge about and understanding of the communication process, the global information environment (GIE) and its potential impact on operations.

1-9. PA personnel must also thoroughly understand the fundamentals of Army operations and the strategic context within which the Army conducts operations. As the GIE compresses the strategic, operational and tactical levels of operations, PA personnel need to appreciate the linkage between public opinion, political decision-making and the national security strategy. They must understand the Army's approach to fighting, influencing events in operations other than war and deterring actions detrimental to national interests.

1-10. PA leaders must, therefore, be among the most informed people in the command. They must be thoroughly aware of all aspects of the operation. They need to know what is being reported about the operation in the global news media and how internal and external audiences are reacting to that information.

1-11. The Public Affairs Officer. The PAO's primary mission is to assess the PA situation, advise the commander on PA issues, assist him in making the best possible decisions, and translate his decision into effective PA operations. PAOs employ the decision-making process to plan, coordinate and supervise the implementation of a PA strategy that helps the commander meet his obligation to communicate with the American public, soldiers, home station communities and the Department of the Army community. PAOs analyze the situation, anticipate issues, assess implications, and develop comprehensive operations, which meet the news and information needs of internal and external audiences and facilitate media operations.

1-12. The Public Affairs Non-Commissioned Officer. The strength of the Public Affairs functional area is its non-commissioned officer corps. PA NCOs are experts on the global information environment, media operations, information strategies, and PA training. They are integral to all facets of the PA planning and decision-making process and provide the essential functional area expertise and continuity required for successful PA operations.

1-13. PA NCOs work closely with the PAO, and in many situations, a PA NCO is the commander's senior PA advisor. Therefore, PA NCOs are fully prepared to assess the PA situation, develop, synchronize and coordinate a PA strategy, implement and monitor PA operations, and measure and evaluate the success of the PA effort.

1-14. The Public Affairs Specialist. In addition to learning traditional soldier skills, PA specialists are trained to support the entire spectrum of PA operations conducted in the global information environment. They are trained on news media operations, news media facilitation, information strategies and information provision. They register media representatives, gather information, develop information products, support news media briefings, respond to news media inquiries and requests for assistance, and track and monitor news media activities throughout the area of operations. They work with news media representatives to gather accurate information and provide timely, balanced coverage of the operation.

1-15. The Department of the Army Public Affairs Civilian. Civilian PA practitioners assigned to Tables of Distribution and Allowances (TDAs) have the same skills as military PA personnel. They provide critical support during war and non-combat operations by providing a vital link between deployed forces and the home station community, and in many situations, may be called upon to deploy with the units they support, or as individual augmentees.

PUBLIC AFFAIRS ELEMENTS

1-16. Battlefield commanders have two sources for tactical PA support. The first is the PA section organic to a warfighting headquarters. The second is the PA unit, which is attached to a headquarters to augment the command's PA capability.

1-17. Conducting PA planning, facilitating news media operations on the battlefield, providing news and information, and executing PA training and support operations is manpower intensive. The austerely staffed PA sections organic to warfighting headquarters will nearly always be overwhelmed trying to meet PA requirements in war and other operations. PA staff sections, therefore, rely on early augmentation by PA units, or individual augmentation if appropriate, to accomplish the battlefield PA mission.

1-18. **Organic Public Affairs Sections.** Organic PA sections are found in warfighting headquarters at various levels including brigades, divisions and corps throughout the Army. Army PA personnel are also assigned to the organic PA sections of joint and combined headquarters.

1-19. In headquarters without organic PA sections, the commander is responsible for PA and must plan and execute PA operations or assign responsibility for PA operations as a special or additional duty to an officer or senior NCO in the command.

1-20. Regardless of the echelon, the PA staff section's primary responsibility is to assist the commander in accomplishing his mission. The staff:

- provides PA information expertise and advice
- conducts PA assessments
- provides analysis of the information environment
- conducts PA planning
- develops information strategies and guidance
- implements PA operations
- measures the effectiveness of the PA effort
- conducts PA training

1-21. The PA staff element controls augmenting PA units. It determines requirements, defines priorities and assigns missions to the augmenting unit. In conjunction with the augmenting unit commander, the staff element task organizes the unit, allocating personnel and equipment to accomplish objectives. If the PA staff element is a Public Affairs Operations Center or a Task Force Headquarters, it will coordinate Armed Forces Radio and Television Service (AFRTS) support activities for the command.

1-22. **Public Affairs Units.** PA units are fully deployable TOE organizations designed to augment the PA staff sections of warfighting units, although they can operate independently in certain limited situations. When a PA unit is deployed to augment a PA staff section, the personnel in the unit cannot be reassigned as replacements or employed as individual fillers for other public affairs elements.

1-23. PA units depend upon the unit they augment for personnel administration, finance, legal and health services, communications, food service, unit maintenance, and supplemental transportation support. PA units operating at corps and below must have the capability to transport all of their TOE equipment in a single lift using authorized organic vehicles.

1-24. There are currently four types of PA units:

- Public Affairs Detachment (PAD)
- Mobile Public Affairs Detachment (MPAD)
- Public Affairs Operations Center (PAOC)
- Broadcast Operations Detachment (BOD)

1-25. **Public Affairs Detachment (SRC 45500AA00).** The smallest of the PA units, the Public Affairs Detachment (PAD) (Figure 1-1) is commanded by a captain and includes seven PA soldiers..

Public Affairs Detachment

CDR	CPT	46A
PA NCO	SSG	46Q
JOURN	SGT	46Q
JOURN(x2)	SPC	46Q
BR JOURN	SPC	46R
JOURN	PFC	46Q
BR JOURN	PFC	46R

Figure 1-1. PAD

1-26. The PAD normally augments a division, separate brigade and armored cavalry regiments and deploys in support of combined, unified or joint operations.

1-27. The PAD commander assumes responsibilities as the PAO or deputy PAO, and the PAD PA personnel are integrated into the supported command's PA section based on operational requirements.

Historical Perspective

The 10th Mountain Division, Fort Drum, N.Y., arrived in Somalia with hundreds of reporters already there, and absolutely no public affairs personnel accompanying them. The first public affairs support arrived at the 2nd Brigade, 10th Mountain Division, 10 days later and only because the JIB dispatched one of its own PADs.

(PA After Action Report, Operation Restore Hope, 10th Mountain Division, Dec 1992)

1-28. Because of the size of the unit, the PAD provides limited:

- Personnel and equipment for digital imagery and audio acquisition.
- Personnel for media escort within the supported unit's area of operations
- Coordination of an information product distribution system that can relay news and information products to members of the supported unit, higher echelons and home station.
- Planning, developing and implementing strategy to support civilian news media and facilitate news gathering efforts throughout the supported unit's area of operations
- Personnel and equipment to coordinate, assist or conduct press conferences and briefings

- Personnel to train, advise and assist leaders and soldiers interacting with or supporting civilian news media within the supported unit's area of operations.

1-29. **Mobile Public Affairs Detachment (SRC 45413A000).** The workhorse of PA units, the Mobile Public Affairs Detachment (MPAD) is a modular, task organizable unit, which provides the full range of PA services. (Figure 1-2). The MPAD is commanded by a major and includes 20 soldiers.

Mobile Public Affairs Det

Command Cell

CDR	MAJ	46A
PA SPVR	MSG	46Z
CLERK	SPC	71L

MEDIA TEAM*

TM OIC	CPT	46A
NCOIC	SFC	46Q
JOURN	SGT	46Q
BR JOURN	SGT	46R
JOURN	PFC	46Q
BR JOURN	PFC	46R

BROADCAST TEAM*

TM OIC	CPT	46A
NCOIC	SFC	46R
BR JOURN	SSG	46R
BR JOURN	SPC	46R
BR JOURN	SPC	46R
BR JOURN	PFC	46R

PRINT TEAM*

TM OIC	CPT	46A
NCOIC	SSG	46Q
JOURN(x2)	SPC	46Q
JOURN	PFC	46Q

*This is one possible configuration. Paragraph 2 members can be configured as one, two or teams of various strengths, based on mission.

Figure 1-2. MPAD

1-30. The MPAD normally augments a Corps PA section or a Public Affairs Operations Center (PAOC). In support of a PAOC, it provides manpower and equipment to establish and operate a media center at Theater Army, TAACOM and Corps.

1-31. It may also be deployed to directly support a joint service task force or non-DoD governmental agency conducting disaster relief, humanitarian assistance, counter drug, peacekeeping, or other contingency operations.

1-32. MPADs in direct support of a gaining command PAO provide acquisition capability for print, audio and video.

1-33. Additional MPADs in direct support can expand the media escort capability of the supported PAO, augment divisions and other elements in theater and provide the PAO with staff augmentation. MPADs may be deployed forward to establish media centers or a sub-JIB in a joint environment.

1-34. MPADs have the capability to:

- Monitor and assess the perceptions of external audiences through access to civilian commercial news sources.
- Conduct assessments of the information environment, to include development of a PA estimate of the situation, as the initial part of operational planning.
- Assist the PAO in operational planning and policy and ground rules for media, coordination for logistical support to PA, and coordination of PA operations with higher and subordinate headquarters.
- Plan and develop information products, which will be produced through contracted services and/or the use of organic equipment and facilities.
- Acquire, produce and transmit information products throughout the theater, between the theater and home station, and between the theater and HQDA.
- Create and disseminate print, photographic, audio and video products for external release directly to civilian media who do not have representatives within the theater of operations. Conduct media facilitation and develop information strategies.
- Prepare commanders, staff personnel and other command members for interviews, press conferences, and similar media interaction.

1-35. **Public Affairs Operations Center (SRC 45423A000).** The Public Affairs Operations Center (PAOC) consists of command, media facilitation and post-production sections (Figure 1-3). It is commanded by a lieutenant colonel and includes 32 soldiers.

1-36. The PAOC normally augments the PA staff section at echelons above division to establish and operate a media center in support of civilian and military media representatives working in the theater.

1-37. The PAOC commander serves as the media center commander but works under the control of the PAO of the supported unit or task force.

1-38. In joint or combined operations, a PAOC serves as the Army element of the joint media operations center.

1-39. When the PAOC functions as the Army element of a joint or combined media center, the PAOC commander works for the joint or combined media center commander.

1-40. For major operations in which there is a significant media interest, the PAOC is augmented by up to three MPADs. The MPADs are either integrated into the main media center operation or tasked to operate subordinate media centers at outlying locations. A PAOC can support up to 100 news media representatives. When augmented by three MPADs, the PAOC can support up to 300 news media representatives.

Public Affairs Operations Center

Command Cell

CDR	LTC	46A
XO	MAJ	46A
PAO	MAJ	46A
CH PA NCO	SGM	46Z
PA NCO	SFC	46Q
SPLY SGT	SGT	92Y
MECHANIC	SPC	63B
ADM SPC	SPC	71L
PERSONNEL	SPC	75B
ADM CLERK	PFC	71L

ESCORT TEAM

CHIEF	CPT	46A
PRESS OFF (x2)	CPT	46A
JOURN (x2)	SGT	46Q
JOURN (x4)	SPC	46Q
JOURN (x4)	PFC	46Q

BRIEFING SECTION

CHIEF	MAJ	46A
BRIEF OFF (x2)	CPT	46A
BR JOURN	SGT	46R
BR JOURN	SPC	46R
BR JOURN (x2)	PFC	46R

A/V PRODUCTION

BR NCO	SSG	46R
PROD SPC	SPC	25V
BR JOURN	SPC	46Q

Figure 1-3. PAOC

1-41. PAOCs are modular, task organizable units having the capability to:

- Implement the theater or corps strategy to support civilian news media and facilitate news gathering efforts in theater.
- Coordinate and provide services to registered civilian news media sponsored by the command.

- Provide the personnel and equipment to coordinate and conduct media support within the theater of operations.
- Provide personnel and equipment to plan and conduct daily news media briefings.
- Provide personnel to train, advise and assist leaders and soldiers interacting with or supporting civilian news media representatives.

1-42. **Broadcast Operations Detachment (SRC 45607A00).** The BOD consists of a command element, two broadcast teams and a maintenance team. (Figure 1-4). It is commanded by a major and includes 26 soldiers.

1-43. The BOD augments a fixed or field expedient AFRTS facility under the control of a senior AFRTS facility commander, or it can establish and operate a separate radio and/or television broadcast facility to support theater level operations.

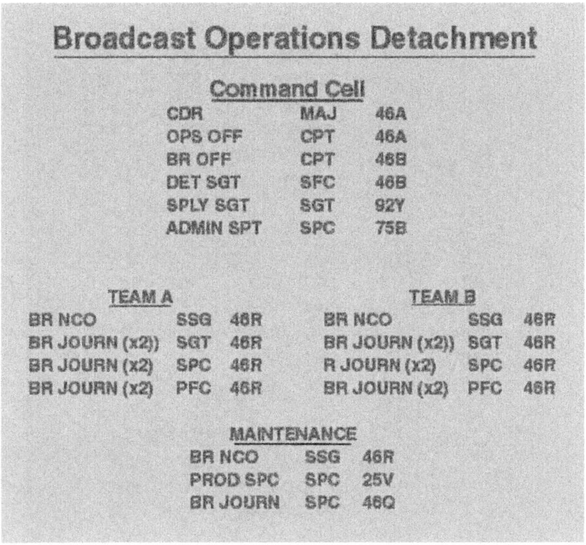

Figure 1-4. BOD

1-44. The BODs must be authorized and equipped by AFRTS to perform this mission. When deployed to perform this mission, the BODs are assigned to the PAOC supporting the command.

1-45. BODs have the capability to:

- Provide on-air broadcasters recorded materials and satellite down links to operate a 24-hour a day radio outlet.

- Provide on-air broadcasters, recorded materials and satellite down links to operate a television station.
- Originate audio and video news, feature and entertainment programming from within theater.
- Provide post production services for audio and video news and feature material supplied in unedited format.
- Provide limited audio and video materials to other public affairs operations for dissemination outside the theater.
- Acquire audio and video electronic newsgathering coverage of operations in the theater for use in internal and external information programs.
- Perform field maintenance and repair above operator level to broadcast equipment organic to the unit.
- Provide the commander with an alternate means of communications when tactical communications are not adequate or not available.

1-46. Public Affairs organizations are built around a force of soldiers who are selected and trained to articulate the goals and missions of the Army.

1-47. More than 65 percent of the total public affairs force and 85 percent of the deployable TOE unit structure is positioned in the U.S. Army Reserve and Army National Guard. These reserve units and personnel must be seamlessly integrated with the active component and focused on supporting the overall Army goals and objectives.

Chapter 2

Public Affairs Functions and Responsibilities

THE CHALLENGE

2-1. The global information environment and continually evolving information communication technologies make it imperative that information and messages be consistent at all levels. The personal comments made by a deployed soldier in a remote area of operations and the official statements released by DoD at the Pentagon must be mutually supporting. The information targeted to internal audiences must parallel the information released through the news media to the American public and other external audiences. The Army's need for security, and the soldier's and family member's right to privacy must be balanced with the Army's obligation to provide timely, accurate, complete information to internal and external audiences. The commander's information strategy must ensure that the information available in the public domain, regardless of the source, does not conflict, contradict or otherwise undermine the credibility of the command or the operation.

Historical Perspective

Civilian news coverage contributed greatly to maintaining soldier morale during Desert Storm. The coverage was generally positive; the American people were behind the operation and soldiers felt this impact. Problems arose when the coverage created rumors, and command information was not consistent with what the soldiers were seeing or hearing in the world media. Family members and non-deployed soldiers were greatly affected by news coverage, often creating problems for rear commanders and detracting from their credibility.

(After Action Report, Desert Storm 1990, Center for Army Lessons Learned)

2-2. Accomplishing this presents unique command and control challenges for commanders, PA practitioners at all levels and others involved in using information to help accomplish the mission in the most effective, efficient manner. It requires careful coordination between staff elements and necessitates continual liaison between levels of command from the tactical through the operational to the strategic.

2-3. Further complicating PA command and control challenges are PA force structure realities. The small size of the PA staff sections organic to war-fighting headquarters necessitates augmentation, especially for operations with a high level of visibility. The availability of augmenting PA units, the majority of which are located in the reserve components, and the difficulties inherent in deploying PA civilians result in heavy dependence on augmentation by individuals. This leads to the creation of ad hoc, unequipped PA elements, which have not trained together or developed relationships with other staff sections or commands, and do not have established internal or external operating procedures.

2-4. For PA personnel therefore, the critical challenge is to rapidly define command and control channels, establish lines of communication and develop operating procedures. The responsibility for doing this usually lies with the Corps PAO who normally leads the commander's PA effort. He identifies requirements, assesses resources and plans, organizes, directs, coordinates and controls the PA operation.

PUBLIC AFFAIRS RESPONSIBILITIES

2-5. Effective PA command and control establishes a public affairs organization based on analysis of mission, enemy, terrain, troops, time available and civilians (METT-TC), tailored to the situation, which reflects the commander's concept of the operation. It ensures that there are sufficient, experienced, PA personnel at each echelon to provide the commander and his force with the most effective and efficient support possible.

2-6. PA command and control begins at the DoD level. The Office of the Assistant to the Secretary of Defense, Public Affairs (OASD(PA)), retains primary responsibility for the development and consistent implementation of DoD information policies and determines who should serve as the initial source of information about operations. Although (OASD(PA)) delegates PA release authority to the combatant commander as soon as practical, it retains responsibility for approving Public Affairs Guidance (PAG), establishes public affairs policy, and coordinates and approves PA strategies and plans.

2-7. The Office of the Chief of Public Affairs (OCPA) for the Army is responsible for Army PA resources. OCPA develops PA doctrine, designs PA organizations, determines training and leader development requirements, identifies materiel needs, and manages PA personnel to ensure that sufficient assets are available, qualified and ready to conduct successful PA operations in support of any assigned mission.

2-8. Commanders supported by their PA staff personnel, plan PA operations for their assigned missions based on the situation, published in DoD directives, instructions, doctrine and guidance, and in coordination with OASD(PA). The CINCs prescribe the chain of command, organize and employ forces, give authoritative direction, assign tasks and designate objectives through component commanders, subordinate unified commanders, commanders of joint task forces and other subordinate commanders. The commander establishes responsive PA structures and ensures that they are provided with the personnel, facilities, equipment, transportation and communications assets necessary to provide adequate PA support. A failure to establish these structures results in a duplication of effort and a waste of resources. The commander is responsible for the full range of PA activities -- PA planning, media facilitation, information strategies and PA training and at sustaining base, community relations. He is also responsible for establishing, resourcing and guiding the operations of Joint Media Operation

Centers and planning all AFRTS radio and television support operations in the area of operations.

Historical Perspective

The establishment of the Joint Information Center under the auspices of the DOT Presidential Task Force on September 1, 1992 was vital to a coordinated and successful Joint Public Affairs effort. The JIC was an "umbrella" organization that served as the clearing house for dissemination of hurricane relief information to the news media. More than 10 federal agencies involved in relief operations had public affairs representatives at the JIC. Daily meetings and consistent interaction among the agencies involved resulted in a coordinated federal information effort.

(Public Affairs Lessons Learned Library, Joint Information Center, Hurricane Andrew, 1992)

2-9. Within the Army, the Corps is usually the hub for PA operations. The Corps commander, supported by his PA staff element, is responsible for the development and coordination of PA strategies, the implementation of information campaigns and the execution of PA operations based on METT-TC, the information environment, and guidance and policy received from the combatant CINC.

2-10. The Corps PAO is the principal PA advisor to the Corps commander and deploys with the lead element of the Corps headquarters. When fully deployed, the Corps PAO section operates from the Corps main command post, with a liaison officer/NCO located at the Corps plans cell. It also is responsible for establishing coordination with the PA elements of higher, lower and adjacent commands.

2-11. The Corps, through the PAO, controls the employment of augmenting Army PA units deployed in support of the operation. Up to one PAOC and six MPADs are normally allocated to augment the Corps PA section, although the size and scope of the mission will determine the actual augmentation required for each operation. The Corps PAO and his staff task organize the personnel and organizations available and allocate the equipment, communications support and facilities. When augmented, the Corps PAO operates a media operations center and establishes satellite centers as required.

2-12. Below the Corps level, the PA staff section organic to a war fighting headquarters is extremely austere. The mission of the PA section below Corps is to advise the commander by providing immediate planning expertise and guidance on issues with critical PA implications. The PA section deploys with the lead elements of the headquarters, and operates from the command's main CP.

2-13. PA units deployed to augment the staff sections organic to a headquarters are normally placed under the control of the supported PAO, who assigns the augmenting PA unit missions and tasks. He will do so in conjunction with the augmenting PA unit commander, who will retain command of his unit and ensure that his unit's personnel are not employed as individual fillers. Whenever possible, augmenting PA units should be linked with the supported command headquarters at that command's home station prior to deployment to facilitate establishment of command and control relationships.

PUBLIC AFFAIRS COMMUNICATIONS

2-14. Reliable, survivable, flexible communications are essential for effective PA command and control. In today's global information environment, information must flow to and from users, up and down the chain of command, and horizontally across the battlefield. Technology has compressed time and space and forward-deployed PA sections can be in direct communication with officials at DoD working PA strategies. The challenge is to ensure coordination and interoperability so that all elements have the communications capability necessary to effectively carry out their assigned mission, especially in today's joint, combined or interagency environment.

2-15. Deliberate, detailed planning can prevent communications shortfalls. PA planners assess their information transmission and reception needs and requirements. They then identify the communications capabilities they need access to, and determine the communications support they will need from command signal organizations. Through close coordination with the staff signal section, the identified PA communications requirements are integrated into the overall communication architecture.

INFORMATION OPERATIONS

2-16. The three central defining characteristics of the global information environment -- the facility of information acquisition and transmission, the speed of information communication and the breadth of information saturation -- combine to increase information availability. The American public, internal audiences, allies and adversaries have ready access to information. Information security is transitory and it is critical that information operations at every echelon are mutually supporting and directed at a clearly defined, decisive and attainable objective.

2-17. Credibility is essential for successful information operations. If an information source is not perceived as believable, then the desired effect of that communication cannot be achieved. Regardless of the source, target or objective of an information effort, in the GIE, credibility is founded in truth and enhanced by validation, corroboration, and consistency.

2-18. Commanders require integrated, coordinated, synchronized information operations. PA operations, which occur at, and impact on, the strategic, operational and tactical levels -- often simultaneously -- are a critical element of these operations. News media coverage of conflicting messages and information communicated by different elements of the command compromises credibility.

2-19. Integrating, coordinating and synchronizing every element of the commander's information operation -- Public Affairs, Psychological Operations, Civil Affairs, Combat Camera, Operations Security and others -- results in a synergistic information strategy. It minimizes the possibility of conflicting messages, which undermine credibility, jeopardize operations and endanger mission accomplishment.

LOGISTICS SUPPORT

2-20. Logistics is critical at all levels of command for Public Affairs mission success, during any phase of combat or garrison operations. Commanders must ensure their Joint Table of Allowances (JTA), Modified Table of Equipment (MTOE), Table of Distribution and Allowances (TDA) or Common Table of Allowances (CTA) reflects appropriate equipment levels to maintain a PA staff and media support under field and garrison conditions. Maintenance also plays an important role in Public Affairs operations. A Public Affairs element that has all its equipment cannot function properly if its equipment is inoperative, broken or deadlined. Each Public Affairs element must develop its own internal SOP in regard to logistics. (See Appendix K.)

2-21. Public Affairs staff members must be trained in the areas of supply, budget, property book, ordering, class A procurement, etc. Public Affairs must be an integral player in all mission and operational planning sessions to ensure logistical requirements are identified and resourced.

2-22. Responsibilities: The Public Affairs staff has the responsibility to identify to its resource manager, property book manager and ordering officer all fiscal and logistical requirements for field operations and home station support. Requisitions for equipment, supplies, services and allowances will be ordered and processed in accordance with appropriate Army Regulations, AR 710-2 *Unit Supply Update* and budgetary guidelines.

2-23. Requesting supplies: Commanders must ensure that equipment and components authorized by JTAs, CTAs, MTOEs, or TDAs are on hand or requested. The organization's supply operation is responsible for identifying, acquiring, accounting, controlling, storing and properly disposing of materiel authorized to conduct the mission of the unit and maintain the soldier. The organization is the foundation of the supply system. Exceptions and procedures are outlined in AR 710-2.

2-24. The Direct Support and General Support Activities provide class 1, 2, 3 (packaged and bulk), 4, 5, 7, 8 and 9 supplies directly to the using units on a customer support basis. These supplies are routinely procured through the unit supply rooms. In the event a Public Affairs element is operationally attached outside its assigned organization, it should coordinate before deployment for logistical support when possible. If prior coordination is not possible, contact for support should be made upon arrival into the theater of operations through the C-4, J-4, G-4, or S-4.

2-25. Accountability: All property acquired by the Army, regardless of source or whether paid for or not, must be accounted for, in accordance with applicable Army regulations and AR 710-2.

- Nonexpendable property is personal property that is consumed in use and that retains its original identity during the period of use. It requires formal property book accountability throughout the life of the item. It will be accounted for at the using unit level using property book procedures. Examples are desks, computers, file cabinets, chairs.
- Expendable items are property which is consumed in use or that loses its identity in use, and all items not consumed in use with a unit price

of less than $100 and not otherwise classified as nonexpendable or durable. It requires no formal accounting after issue to the user. The following classes or types of property will be classified as expendable.

- Supplies consumed in the maintenance and upkeep of the public service. Examples are oil, paint, fuel and cleaning and preserving materials.
- Supplies that lose their identity when used to repair or complete other items. Examples are assemblies, repair parts, and accessories.
- Office supplies and equipment (such as paper, staplers and hole punchers) with a unit price of less than $100.
- Durable property is personal property that is not consumed in use, does not require property book accountability, but because of its unique characteristics requires hand receipt control when issued to the user. Examples are hammers, lawnmowers, audiovisual production material and books.

2-26. Conservation of resources and property accountability is ultimately a supervisory responsibility. Property responsibility must be assigned and acknowledged in writing using hand receipts and property books as outlined in AR 735-5, *Policies and Procedures for Property Accountability* and AR 710-2.

2-27. Property book: Effective supply support at the using element or property book level requires timely and accurate processing of supply requests and receipts, accurate accounting records and adequate property control. Turn in, transfer, substitutions, hand receipt, etc., are accomplished in accordance with Army Regulations, AR 710-2 and logistical SOPs.

2-28. Budget: Budgets must be programmed in advanced. Organizations plan their budget in the previous fiscal year They must be established and managed with the appropriate command resource manager/budget analyst. When allocating funds consideration must be given to equipment replacement and upgrades, recurring supply needs TDYs, maintenance, contracts, etc. Normal operating funds are allocated/dispersed by a public affairs element operational headquarters; however, during deployments for exercises/operations funds may be available from the tasking headquarters up front or on a recuperative basis.

2-29. Maintenance: Public Affairs elements must maintain their equipment in a deployment ready state. Preventive Maintenance Checks and Services (PMCS) is an important part of the maintenance program, and is the user's responsibility. The Public Affairs element's operational headquarters provides maintenance support. For example, an embedded Public Affairs section assigned to the Headquarters Company of a separate brigade would seek maintenance support from the Headquarters Company, then the brigade maintenance section. The company/brigade's maintenance SOP would be followed for 1st, 2nd, 3rd, etc., echelons of maintenance. Maintenance for communications and data processing (computers) equipment is coordinated through the G6 (DOIM).

2-30. Local Purchase: Local purchases may be an option for procurement provided the action is in the best interest of the government in terms of timeliness, quality and cost. Local purchase requests must be made in accordance with AR 710-2. Approval for local purchase of nondevelopmental items starts at the first level of command authority and is accomplished in accordance with local policies and Army regulations. Nondevelopmental items is a generic term that covers materiel available from a variety of sources with little or no development effort from the Army. Sources include commercial items which fully meet an approved need, items being used by other U.S. services or agencies or items used by military or other agencies of foreign government. Most of these purchases are covered under the IMPAC Card program. Local guidance covers implementation.

2-31. Class A Procurement/Credit Card: Class A agents and government credit card holders must be identified and trained prior to their ability to accomplish those functions. Training is routinely accomplished at the installation level. Purchases for other than national-stock-numbered items are routinely accomplished using Class A agents and ordering officers, and the U.S. Government Credit Card.

2-32. SSSC: Self Service Supply Centers are managed at the installation or theater level. Users are required to have a valid SSSC account. Accounts are identified by DODAAC or UIC. Subaccounts are authorized IAW AR 710-2. Field resupply centers are often established at divisional-level logistics bases.

2-33. The best means of ensuring supply discipline is to be proactive and not reactive in supply operations. Enforcing compliance with regulations requires constant emphasis.

Chapter 3

Public Affairs Planning

During Operation Just Cause, PA planning and integration were inadequate. Commanders at all levels failed to involve public affairs officers in planning from fear of OPSEC leaks. The results were insufficient PA guidance provided to soldiers, family members and commanders; sometimes miscommunication to and confusion within Army family elements; and misuse of PA assets. A matter of urgent concern was the failure to plan for and use Reserve Component PA assets to relieve the pressure on an already small active PA force so that it could better handle both internal and external communication.

- Public Affairs After Action Report, TRADOC LLC, Phase II, Sept. 90

COMMUNICATING INFORMATION

3-1. Public Affairs operations assist the commander in communicating information and messages about his force and the operation to internal and external audiences. Like other operations, PA operations are conducted to bring about clearly specified, meaningful objectives, which support the commander's intent and contribute to mission success. Those objectives are defined in terms of the effect the PA operations are intended to have on target audiences -- the impact on target audience behavior that is desired -- and are measurable.

3-2. Once PA objectives are defined, PA operations are planned and executed to achieve those objectives. PA operations focus on the communication process -- an on going, dynamic, ever-changing process. The communication process is composed of elements involved in receiving, collecting, analyzing and interpreting data, identifying and analyzing audiences and formulating and transferring messages. This process is used to bring about a specified objective, while measuring and analyzing the outcome and effectiveness of the effort.

3-3. To support the commander's effort to communicate, PA professionals concentrate on five basic functions or core processes — planning, media facilitation, information provision, force training and community relations.

3-4. This chapter focuses on the systematic process for Public Affairs planning and decision-making. It addresses the information environment and the impact of information at the strategic, operational and tactical levels across the range of operations that requires public affairs considerations be totally integrated into the planning and decision-making process. Doing so enables PA personnel to prepare for potential situations, to synchronize efforts with other agencies that manage information communication, and to more successfully influence the coverage, interpretation and understanding of events. It limits the need for reactive, defensive attempts to buy time or control damage.

3-5. PA planning prepared in support of the CINC's theater campaign plan requires a series of decisions related to policy at the national level and the techniques at the tactical level. From policy to techniques, however, basic planning considerations are the same: What should the PA objective accomplish? With what audience? When? How? PA planning must not only be done at all echelons and within national policy but also within the limits of operational plans and capabilities.

TYPES OF PLANS

3-6. The amount of time available significantly influences the planning process. Two different methods of planning are described in the JCS-published Joint Operations Planning and Execution System (JOPES).

3-7. **Deliberate or Peacetime Planning** is the process used when time permits the total participation of commanders and staffs. Development of the plan, coordination among supporting commanders and agencies, reviews by staffs, planning conferences and development of proposed public affairs guidance can take many months. Deliberate or Peacetime Plans are prepared in prescribed formats--the complete operational plan (OPLAN) or the conceptual operational plan (CONPLAN).

3-8. **Time-Sensitive or Crisis Action Planning** (CAP) is conducted in response to crisis where U.S. interests are threatened and a military response is being considered. Crisis Action Planning is carried out in response to specific situations as they occur and that often develop very rapidly.

3-9. It is within the CAP process that established, working relationships between the PAO and operational planning staffs are crucial to the inclusion of PA considerations into OPLANS and OPORDS.

3-10. Both deliberate and crisis action planning are conducted within JOPES. Joint Pub 5-03.2, JOPES Volume II, describes detailed administrative and format requirements for documenting the annexes, appendixes, etc. of operational plans, and conceptual plans, the products of deliberate planning.

3-11. The purpose of JOPES is to bring both deliberate and crisis action planning into a single architecture to reduce the time required to complete deliberate or crisis action planning. This makes the refined results more readily accessible to planners, and makes it a more manageable plan during execution.

3-12. The overall procedures are the same, at all echelons, for both deliberate and crisis action planning.

- Receive and analyze the task to be accomplished
- Review the situation and begin to collect necessary intelligence
- Develop and compare alternative courses of action
- Select the best alternative
- Develop and get approval for its concept
- Prepare a plan
- Document the plan

Operation Just Cause

The basic problem---planning. No discussion concerning use of personnel can proceed without an understanding of the planning problem. Public Affairs in general was not sufficiently planned for by leaders or public affairs officers for Operation Just Cause. PAOs were not given time to plan. Only outstanding unit mission accomplishment, American public support and the hard work of public affairs personnel, prevented major PA failings in Panama. A longer duration and less popular action could have turned into a public affairs disaster.

The SOUTHCOM, XVIII Airborne Corps, and Army Special Operations Command PAOs were not informed that the operation would occur until 17 December and then they were given instructions not to discuss it with key persons on their staff. The 82d PAO did not learn of the operation until 18 December, 7th ID PAO, 19 December and OCPA **one hour** before H hour.

More critical however, was the absence of joint coordination. OASD-PA received a PA plan from SOUTHCOM public affairs in November, but the plan was never coordinated due to worries concerning possible security leaks. The XVIII Airborne Corps PAO indicated that he also knew of the operation in general terms in November but was unable to coordinate planning with the Director of Public Affairs SOUTHCOM. Because the plan was not staffed, OCPA was caught unaware. Divisional public affairs officers all indicated that they were not sufficiently drawn into planning. Some PAOs said they never saw a plan. It is obvious that sufficient public affairs planning did not occur **at any level**.

There was also a ripple effect downward caused by the lack of staffing by the Joint/OASD-PA. In addition, senior leaders at division level and above did not draw PA officers into planning to maximize the limited planning time that was available. Commanders strongly complained about poor balance of coverage in the media, inability to send command information at the same pace as civilian media reporting, and lack of sufficient public affairs guidance; yet senior leaders are reluctant to draw PAOs into the planning process to prevent these problems from occurring. Army leaders must come to grips with this dilemma.

(Public Affairs After Action Report TRADOC Lessons Learned Collection Phase II. Sept 1990)

3-13. An OPLAN is a complete, detailed plan. It includes a description of the concept of operations from the commander's perspective and presents additional annexes provided by various staff sections which identify specific functional area requirements, restrictions, limitations, or considerations. The inclusion of a public affairs annex is essential to successful integration of PA principles and guidance into the OPLAN.

3-14. A CONPLAN is an abbreviated operational plan, which requires considerable expansion or alteration to convert it into an OPLAN or OPORD. Detailed support requirements are not included. The commander determines what annexes will be included to complete the CONPLAN.

3-15. A Public Affairs Estimate is an assessment of a specific mission from a Public Affairs perspective. It is an examination of critical Public Affairs factors, their influence on the planning and execution of operations, and their potential impact on mission success. The senior PAO at each echelon is responsible for consolidating information and preparing the PA Estimate. A sample PA Estimate is included at Appendix C.

3-16. The Public Affairs Annexes to OPLANS or CONPLANs provide the details and instructions necessary to implement Public Affairs media facilitation, news and information provision, and force training operations. It is coordinated with all staff agencies, especially those that significantly impact the information environment -- Psychological Operations, Civil Affairs, Signal, and Military Intelligence -- to ensure that Public Affairs activities are synchronized with other activities.

DELIBERATE PLANNING

3-17. A commander continually faces situations involving uncertainties, questionable or incomplete data, or several possible alternatives. As the primary decision maker, the commander, with the assistance of the staff, must not only decide what to do and how to do it, the commander must also recognize if and when to act. How the commander arrives at a decision is a matter of personal determination. However, superior decisions (those, which offer the best solution, decisively, at precisely the correct time,) result from the commander's thorough, clear, and unemotional analysis of facts and supported assumptions. This is done through the "deliberate planning process."

3-18. To support the commander's decisions and command objectives, the PAO must develop a thorough, clear, comprehensive public affairs strategy. This strategy allows the PA to link public affairs considerations into planning for contingency, future and current operations. With the PA strategy, the PAO defines the public affairs perspective of the operation, and identifies how the Army public affairs involvement in this operation supports strategic goals. It provides the intent for PA operations and the Army approach to meeting the information needs of critical internal and external audiences. It is the framework for defining and developing the PA scheme of operations.

3-19. Based on the PA strategy, PA plans are developed for integration into OPLANs. A PA plan is produced by the operational commander's PAO, and it details the media facilitation, news and information provision, and force training and support procedures which will be employed in support of the operation.

3-20. The first crucial step in fulfilling the PA strategy requires the PA Plans officer/NCO to establish and maintain a routine, ongoing relationship with operational planners within the organization. The PA plan is coordinated with key staff agencies, integrated into the OPLAN as a PA Annex. Synchronization with these other activities ensures services and support required by the PAO is provided and multiplies the impact of the PA plan. This process is followed at subordinate echelons as planning guidance is communicated down the operational chain of command.

3-21. There are five phases in the deliberate planning process. Items in parenthesis identify PA actions performed within each phase:

- **Phase I** - Initiation. The task assigning directive outlines the major combat forces available for planning, gives general planning instructions, lists assumptions for planning, and specifies the product document such as an OPLAN, CONPLAN. (PA planners begin assessing the information environment, its impact on operations and the PA requirements to operate within a specific arena.)

- **Phase II** - Concept Development. (Using the supported CINC's mission statement and concept of envisioned operations, the supporting PA planners analyze the mission, formulate tentative courses of actions and develop the PA Estimate for the operational scenario and requirements.)
- **Phase III** - Plan Development. Subordinate commanders use the CINC's concept and the allocated major combat forces as the basis to determine the necessary support, including forces and sustaining supplies for the operation. (The PA planners provide the CINC with recommendations for public affairs assets required, phasing of PA forces and support into the theater of operations, and perform a transportation analysis of their movement to the destination to ensure that the PA segment of the entire plan can feasibly be executed as envisioned. For the supported CINC's PA requirements, above those organic to the tasked major combat elements, the supporting commands [force providers] of each service, as much as possible, identify real-world PA assets to take part in the plan and sustainment to meet requirements. The supporting command identifies PA requirements in OPLANs, OPORDs, and taskings, through operational channels, to major subordinate commands.)
- **Phase IV** - Plan Review. The review process is more than a single phase in deliberate planning. The Joint Staff performs or coordinates a final review of operations plans submitted by the combatant CINCs. It is a formal review of the entire operation plan. Approval of the plan is the signal to subordinate and supporting commands to develop their plans in support of the CINC's concept. (PA planners do not wait until the plan is approved before beginning to develop their supporting plans; they have been involved in doing this, while coordinating with their command's planning staff. In the meantime the CINC has been building the overall plan.)
- **Phase V** - Supporting Plans. The emphasis in the Supporting Plans Phase shifts to the subordinate and supporting commanders. (This is the phase in which PA planners begin to concentrate on how to meet tasks identified in the approved operation plan by preparing Public Affairs Annexes to supporting plans. This input outlines the actions and relationships of assigned and augmenting PA assets.)

INFORMATION ENVIRONMENT ASSESSMENT AND PA ESTIMATE

3-22. Planning fosters effective application of knowledge, logic, and judgment. Analysis of the information environment (IE) starts the process used to develop an estimate of the situation.

3-23. The IE analysis provides the basis for the development of all PA operational plans and is a channel for integration of strategic, operational and tactical planning guidance. It is a method of identifying factors within the information environment that have potential implications for the planning and execution of Army operations. PA planners study and evaluate the dynamics of the area information environment to identify specific public affairs operational considerations.

3-24. Analysis of the information environment focuses on research into the following areas, which will be put into the "Situation and Considerations" portion of the PA estimate:

- Information infrastructure
- Media presence
- Media capabilities
- Media content analysis
- Public opinion assessment
- Information needs assessment
- Impact assessment/courses of action (COA)

3-25. An analysis of the IE using this approach builds a complete picture of the conditions facing commanders and their PA forces, providing them the tools necessary to anticipate trends, actions, issues, and conflicts. The PA staff officer or NCO conducts research and assessment for the estimate then evaluates, prioritizes, and suggests courses of action that public affairs can best support, while considering the information environment. To acquire all the information necessary for an accurate picture of the operational environment, the staff officer or NCO must work closely with intelligence, civil affairs, psychological operations, military police, visual information and other staff sections involved with information gathering.

3-26. The PA assessment must include those aspects under the control of the commander, as well as those the commander cannot control. This can only be achieved with a thorough integration of PA planning at all stages and into all aspects of the planning and decision-making process. Although a variety of techniques may be used in the analysis of the IE, the PA assessment should address the following primary categories.

- **Information Channels and Infrastructure**. This element focuses on an assessment of the information infrastructure. It addresses the resources, communications facilities, organizations, and official and unofficial information channels available within the area of responsibility (AOR). It addresses the means to transmit and receive unofficial information. It addresses specific requirements for American Forces Radio Television Service (AFRTS) information services and the availability of assets to meet theater requirements. It identifies the availability of host nation telephone service for voice and data transmission, the accessibility of audio/video channels, the prevalence of private communications devices such as cellular telephones, facsimiles, computers with modems, radios and televisions, and the nature of the information available through these information channels. It addresses alternate means of voice and data communications, whether military or government contracted, for use in the absence of host nation information channels and infrastructure. Much of the information required for this category may be obtained through civil affairs or psychological operations elements assigned or attached to the command and U.S.Information Service offices supporting consulates or the embassy within the area of operations.

- **Media Presence**. This is an assessment of the media presence in the area of operations prior to the introduction of American forces and an assessment of the expected level of media presence commanders should anticipate once deployment begins. It includes a description of the type of media (print or broadcast), the visibility of the media (local, national, or international; American or foreign), and the focus of the news media present (news or entertainment) covering the operation. The assessment of the media presence should address the authority under which media representatives are operating (open or closed borders, and free press or controlled press) and the reporters' degree of access to the theater of operations.
- **Media Capabilities**. This element is an assessment of the media's information collection, production, transmission and communication capabilities in the AOR. This element analyzes the technological capabilities of the media representatives present within the AOR. It describes their level of sophistication (if they must transport products out of the area of operations for transmission to parent media or do they have self-contained interactive satellite telecommunications access). It also addresses the media's level of logistics support and its potential impact on Army commanders who are required to provide the media free and open access to the AOR. It includes information about their transportation assets, resupply channels, and equipment maintenance requirements. Additionally, the media's general ability to provide their own security should be assessed.
- **Media Content Analysis**. Media content analysis is an assessment of news coverage, the media's agendas and an analysis and prioritization of the potential strategic and operational issues confronting the command. Media content analysis assesses what is being said, by whom, and how it is being presented. It is a constant process that must begin well before planning for a specific operation begins and continues through Mobilization, Deployment, Employment, Sustainment, and Re-deployment. Content analysis reveals the meaning, tone, and accuracy of messages, how the information was presented, and the cumulative affect of the information. A media content analysis will provide an evaluation of the quantity of coverage, both in and out of theater, and the nature of that coverage. This will assist the commander to understand the strategic context, the measure of success and the definition of an end-state for the operation as viewed from outside the command and the Army itself. It will also be an essential element of friendly information (EEFI), as explained in FM 3-13 (100-6), in determining objectives and strategies for communicating the Army perspective, and for working to achieve a balanced, fair and credible flow of information.
- The specific methods for conducting a media content analysis are explained in Appendix O.
- **Public Opinion**. A public opinion assessment surveys the national and international attitude about the operation and the command, leaders and soldiers conducting it. This assessment looks at the perceptions held by major audience and coalition groups, and the relative solidity or strength of those attitudes. It addresses the perceptions held by international

audiences: those traditionally allied with the United States and those traditionally considered to be adversaries of the US. The public opinion assessment should include as a minimum, consideration of the following groups:

- American public (general)
- Civilian political leadership
- Coalition and allied forces and their general population
- Host nation citizens
- International public
- Internal command audience
- Home station community
- Specific special interest groups (if needed)
    - In determining the effects of the media on public opinion, there are three general types of evidence which explain behavior response: direct indicators; indirect indicators; and post-event sampling. Direct indicators are evidence that provide a direct link between the information received by the public and the behavioral response. These indicators include but are not limited to: personal interviews and surveys to estimate awareness and understanding of an issue; dissident group marches, meetings, advertising and other activities; monitoring internal and external law, order and discipline activity; and chain of command after action reports, staff journals and duty logs. Indirect indicators are evidence that identifies behavioral response generated by separate events or activities which appear to be the result of reception of media information. These indicators include: cause-effect estimates from information products and sources other than the military or civilian commercial media; interest level in news media products; shifts in social or economic trends; shifts in political support.
    - Post-event sampling considers the qualitative and quantitative statistical evidence that identifies the level of and nature of awareness and behavioral response to information. This includes the results of surveys, interviews, group observation, probability and non-probability samples, which will identify if and how the public was influenced by information products or messages.

- Information Needs. This is an assessment of the information needs and requirements of the previously identified key publics. It analyzes and prioritizes key external and internal audiences and assesses their news and information expectations. It identifies the types of information that should be made available to soldiers, their family members, other home station community audiences, the American public, and the host nation local populace. It will identify other audiences, such as allied or adversary leaders and publics that will be interested in available "cross-border" information.

PA ESTIMATE AND PA GUIDANCE COORDINATION

3-27. The purpose of the PA Estimate is to determine whether the mission can be accomplished and to determine which COA can best be supported by public affairs. In preparing its estimate, the Public Affairs staff:

- Reviews the overall mission and situation from the public affairs and information environment perspective.
- Examines all public affairs factors impacting on or impacted by the mission.
- Analyzes each COA from the public affairs perspective.
- Compares each COA based on the public affairs functional analysis.
- Concludes whether the mission can be supported by public affairs, and from the public affairs perspective, which COA can best be supported.

3-28. The Public Affairs Estimate summarizes the information environment, prioritizes the major issues confronting the command and predicts anticipated outcomes in detail. It measures the effectiveness of previous and current information strategies, and based on this evaluation, identifies possible courses of action to support command PA objectives. The PA Estimate also contributes to the development of Public Affairs Guidance (PAG) for specific operations or missions. PAG is a primary tool that guides commanders and PA leaders in the application of doctrine and policy during operations. PAG provides the PA force at all echelons standard operating procedures.

3-29. But to be effective, PAG must be developed with the needs of the front-line PA force in mind. PA planners must be able to "see" and "feel" the battlefield. They must have an understanding of the information environment and how it will change throughout the operational continuum. They must be aware that all the resources available at the planning headquarters may not be available or feasible in the theater of operations. Issues that need to be addressed include information release authority restrictions (national, theater or local). These restrictions often place the PA leader in a difficult situation -- one in which an overwhelming number of news media on the scene will seek answers to legitimate questions about unfolding events -- activities that the PA leader cannot discuss. The result is a loss of credibility for the Army.

3-30. DOD policy requires that proposed PAG be provided to the Assistant to the Secretary of Defense for Public Affairs (OTASD-PA) by the unified, specified and other major commands for all operations. This requirement includes major joint training exercises that could attract national and international attention. Subordinate command PA leaders should conduct continuing PA assessments as a part of mission training for theater-specific contingencies in anticipation of PAG development requirements there.

PAG DEVELOPMENT

3-31. Upon receipt of a "warning order," the commander, through the PA staff, will begin development of proposed PAG. In reality this warning order may be preceded by a telephonic "heads up" call from a PA planner at a higher headquarters that allows PA planners to begin working on proposed PAG before the hard copy of the warning order arrives. This proposed PAG should be based on the warning order or other planning guidance, the proposed command operations plan (OPLAN), and the PA Estimate. Once the proposed PAG is developed, it is staffed through command staff. Once approved by the commander, it is forwarded through major command and Unified/Specified command PA channels to DoD.

3-32. Commanders of Unified/Specified commands should ensure the proposed PAG is coordinated with appropriate elements and functional staffs within the theater of operations. This includes governmental and non-governmental organizations such as the State Department and its embassies, civil affairs, country assessment teams, host governments, allied force public affairs teams, the four U.S.military services and their subordinate commands.

3-33. PA leaders at all levels, specifically major command and above, should work to approve PAG as quickly as possible in order to provide subordinate PA leaders the opportunity to develop and implement PA strategies to support their commands. The format for PAG is included as an appendix to this manual at Appendix E.

PA ANNEX DEVELOPMENT

3-34. Once the PA estimate and proposed PAG are completed, and the other staff officers have completed their estimates, the commander selects a course of action. The commander then outlines it to the staff. The commander may select one of the proposed COAs, a combination of two or more, or a completely new one. The PA staff must then be prepared to enter the plan development phase (Phase III) which requires development of a PA annex. A format for a PA Annex is included at Appendix D.

3-35. The operation-specific approach to conducting public affairs activities is called a PA scheme of maneuver. This PA scheme summarizes the commander's PA intentions, and details the media facilitation, news and information provision, and force training and support procedures, which will be employed to support a particular operation.

3-36. The PA scheme consists of the PA estimate of the situation, higher command PA guidance, and the selected course of action. It is coordinated with key staff agencies, integrated into the operation plan through the development of a PA annex, and synchronized with the other activities to be executed as part of the basic plan.

3-37. The PA scheme, when included in the PA annex, should not only identify public affairs force requirements for the operation, but more importantly, it must provide the commander a visual picture of how public affairs will support the commander's concept of operation as outlined in the plan.

3-38. The PA activities addressed in the PA scheme of maneuver are:

- **Media Facilitation**. Media Facilitation is activities executed to support news media efforts to cover the operation, facilitate the timely, accurate, balanced provision of information which communicates the Army perspective, and minimizes the media disruption of operations or endangerment of mission accomplishment. Media facilitation is accomplished by the early establishment of a media center as the focal point for media representatives seeking to cover the operation. Normal media center operations include scheduling briefings, coordination for interviews; responding to media queries; coordinating unit visits and media escort requirements; and resolving media - military incidents. To prepare for encounters with the media, commanders must accept and understand the role of the news organizations and the journalists in the theater, and their capabilities in getting information from the battlefield or area of operations. Commanders must provide media access to the force, keeping in mind the impact their technology will have on operations security. Commanders must identify and provide support and resources to assist the media in their mission.

- **Information Strategies**. Activities executed to fill the news and information needs and expectations of internal and external audiences. Proliferation of personal computers, the World Wide Web, the Internet, on-line services, fax machines, E-mail, cable television, direct broadcast satellites, copy machines, cellular and wireless communication and many other information technologies have created an endless stream of data and information that flow into a world filled with images, symbols, words, and sounds. Public affairs specialists acquire information using a variety of sources. Because of the volume of information and the vast number of potential distribution mediums, the PA staff uses a systematic acquisition strategy. They acquire information from participants, leaders, developed sources, the media, research and development, intelligence, culture at large, and subject matter experts. Print, video, audio and electronic information products are provided to deployed soldiers, home station audiences such as family members and the home station community and news media representatives using contracted services and organic military assets. They communicate the Army perspective and contribute to timely, balanced coverage of the operation.
- **Force Training and Support**. Activities executed to assist members of the DA community in interacting with media representatives. Force training and support are conducted to educate soldiers, family members and DA civilian employees on their rights and responsibilities with respect to news media representatives attempting to provide coverage of an operation and related issues. It focuses on helping them to respond when they encounter news media representatives seeking interviews, photo opportunities, responses, reactions, interpretations or comments on an operation, policies or events. The intent of force training and support is to assist members of the community and media representatives in approaching each other with mutual respect. Training for public affairs personnel expands on soldier and unit leader training. It stresses individual as well as collective tasks with an aim of developing units fully prepared to accomplish the range of public affairs missions. It integrates public affairs into the battle staff and trains PA planners to assess the operation environment from a public affairs perspective, produce a PA Estimate, develop the PA Annex and PA Guidance.

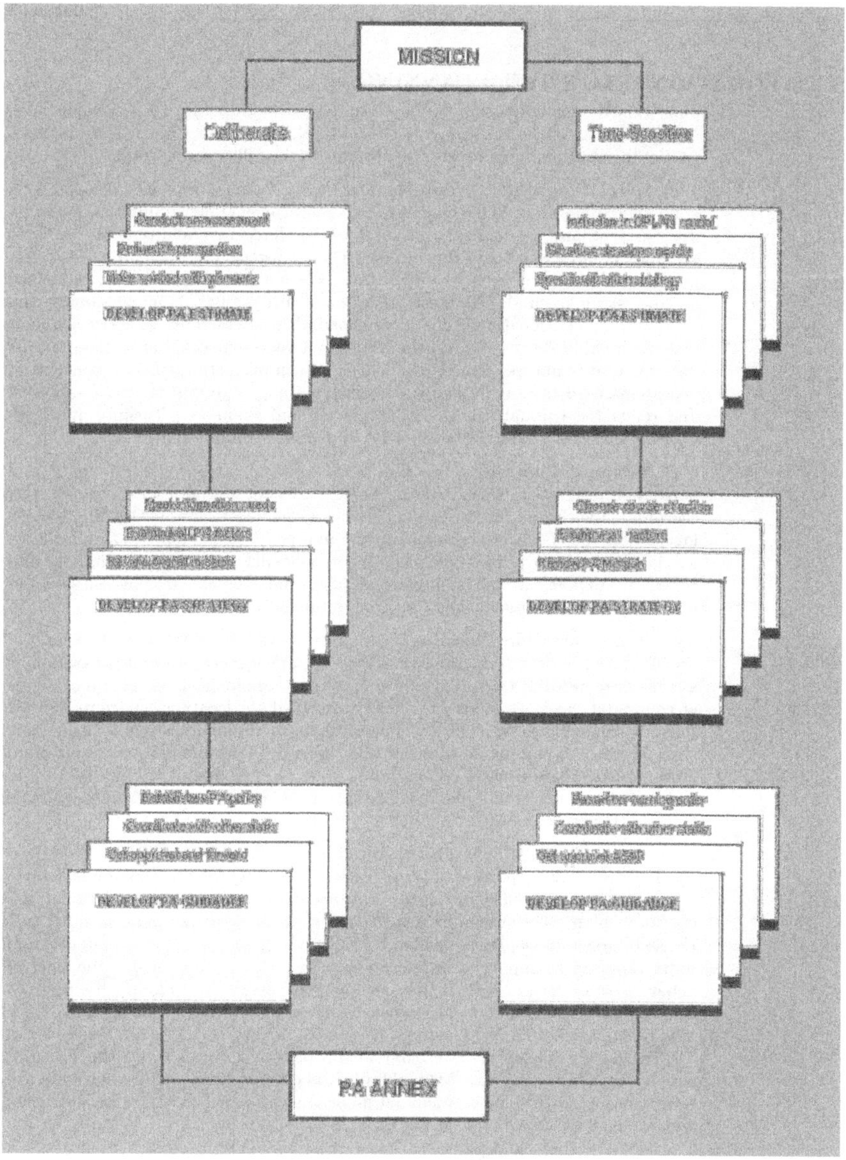

Figure 3-1. PA Planning

PA INTEGRATION INTO STAFF PLANNING

3-39. Concurrent with formulation of the PA staff estimate, PAG development, and production of the PA Annex to the OPLAN, PA planners must be an integral part of the staff planning process, especially on the following matters.

3-40. **Force Planning**. In force planning, the PA staff works with the J3/G3/S3 staff. Force planning consists of PA force requirements determination, force list development and refinements in light of PA force availability and PA force shortfall identification and resolution. In force list development, the PA assets needed to meet the mission are identified. Force availability is considered based on the strength and readiness of organic PA units, their personnel and equipment. Identification of PA force shortfalls addresses the lack of organic or mission-capable PA assets and the additional requirements and augmentations for PA units and personnel needed to accomplish the CINC's concept of operations. All taskings for unit or personnel augmentation must be validated and requested through the J3/G3/S3 operational channels. Tasking authority does not exist PA staff to PA staff or PA staff to subordinate unit.

3-41. **Support Planning.** To plan for logistical support of PA units and personnel assigned to carry out the CINC's concept of operations, the PA staff coordinates with and identifies support requirements to the J4/G4/S4. Specific logistical areas of concern include support in all classes of supply to the PA force, availability and authorized levels of support to civilian media, local purchase and contract support, property accountability, and vehicle transportation and maintenance support as tasked for through the J3/G3/S3.

3-42. **Transportation Planning**. PA forces move from their home station to a specified destination in the theater, either as part of their parent organization or a task-configured PA unit. This movement involves planning by several echelons of command, possibly stops at several intermediate locations en route, and a schedule constrained by a variety of operational requirements and priorities. Key staff for the PA planner to interact with include the command's transportation officer, movements control officer, and staff officers within the J3/G3/S3 and J4/G4/S4 that have staff supervision in this area. Key items PA planners need to track in this process are:

3-43. **Time-Phased Force and Deployment Data (TPFDD).** The TPFDD is the JOPES database portion of an operations plan. It contains time-phased force data, non-unit related cargo and personnel data, and movement data for the operation plan. The Appendix 1 to Annex A of the operation plan is the Time-Phased Force and Deployment List (TPFDL) which identifies types and/or actual units required to support the operation plan and indicates origin and port of debarkation or ocean area. It may also be generated as a computer listing from the TPFDD. PA planners must ensure that the TPFDD/TPFDL contains their unit line numbers (ULNs) for units, personnel, or cargo. Assets not listed on the TPFDD do not deploy. It is crucial to the planning process that the PA staff closely coordinate with the J3/G3/S3 and J4/G4/S4 to ensure that PA assets are reflected on the TPFDD or included as organic/attached assets to parent units with validated ULNs on the TPFDD.

3-44. **Destination (DEST)** - the geographic location where the force is to be deployed/employed.

3-45. The distances between the port of debarkation (POD) within the theater of operations to the destination (DEST), to the port of support (POS), to the marshaling area or assembly area. Where troops land at the APOD (Aerial Port of Debarkation) or SPOD (Seaport of Debarkation) they may be substantial distances from the port where the PA element's equipment arrives in theater and operations begin.

3-46. Transport of equipment must be planned for it to be available at the earliest possible date within the theater of operations. Thus, the PA planner must set a realistic, achievable required delivery date (RDD). This must be aligned with the CINC's required date (CRD). Planners begin with the RDD to establish two interim dates, the earliest arrival date (EAD) and the latest arrival date (LAD). Once these dates are established, then the ready to load date (RLD) and the available to load date (ALD) are established at home station to meet the earliest departure date (EDD).

3-47. **Communications/Automation Planning**. Key planners on the coordinating staff for communications and automation planning are the J6/G6, G3/S3 and CE officer. Specific concerns include priorities for radio/telephone communications, satellite uplinks and downlinks, number of telephone links/trunks allocated to PA requirements, E-mail access, and inclusion into the Communications Electronics Operating Instructions (CEOI). In addition, consider possible development of web pages or sites, like BosniaLink, the Task Force Eagle Homepage, the Desert Voice in Kuwait or Task Force Falcon in Kosovo.

3-48. **Information Environment**. When formulating PA plans and coordinating on the overall plan with the commander's staff, PA planners work closely with the staff element having supervising responsibility for each separate issue.

- Operational Security - G3/S3
- Psychological Operations - G3/S3
- Civil Military Operations - G5/S3
- Combat Camera Operations - G3/S3
- Armed Forces Radio and Television Operations - G3 and Armed Forces Information Service

APPLYING METT-TC

3-49. To function as part of a deployed or deployable organization the PAO and PA NCO must think and state requirements in terms that the rest of the organization can understand. Moreover, the PAO must fit the operational PA requirements into the operational planning procedure of the organization about to deploy. Operators think in terms of mission, enemy, terrain and weather, troops, time available and civilian considerations or METT-TC. METT-TC is used to envision how the operation will occur, to identify potential risks or hazards, and to define troop and equipment requirements.

3-50. **Mission** - alert, marshall, deploy, operate (internal information, media facilitation, information provision) redeploy

3-51. **Enemy** - rumors, disinformation, propaganda, OPSEC

3-52. **Terrain and weather** - theater of operation, theater of war, area of responsibility, intermediate staging base, homestation and weather condition

3-53. **Troops** - embedded assets, units (PADs, MPADs, BODs, PAOCs) AFRTS, HTNRs, ABS, NBS, JIBs, Star and Stripes, stringers, surrogate PAOs, Adjutants, and S1s. All other AC services, signal, USIS, DoS and homestation forces and audiences

3-54. **Time available** - timeline, transition, reports

3-55. Essentially, the information required for a METT-TC analysis is provided by the PA estimate of the situation, which contains the selected course of action and detailed descriptions of PA actions to be performed. These requirements are then translated into the command's planning language and format, resulting in the PA Annex to the OPLAN or OPORD.

3-56. At theater level and above, the PA annex is normally Annex F to the OPLAN. At corps and below, commanders can tailor their plans to fit specific needs or preferences, so the PA annex may fall in another location among the annexes. Regardless of where it is located, the PA Annex is used to provide information about the conduct and execution of public affairs operations in support of the basic OPLAN. The PA annex outlines the situation, identifies the specific PA mission and explains the concept of the operation. It also provides detailed information and guidance PA personnel need to conduct successful PA operations at the operator level. A sample PA Annex format is included in this manual at Appendix D.

3-57. Phase IV, Plan Review, consists of staff coordination and plan adjustment or correction.

3-58. The final phase of the planning process, Phase V, Supporting Plans, follows the same course as the first three phases, with attention aimed at the specific aspects of the overall plan. These supporting plans focus on conducting specific operations, which must be successful in order to guarantee success of the larger mission. PA support to these supporting plans is as important as PA coordination and input to the main campaign plan.

STANDING OPERATING PROCEDURES

3-59. Integral to the operational effectiveness of PA sections are their standing combat operating procedures. These routine procedures ensure that all members of the section are working in concert toward the same PA objectives and that PA activities are easily blended into the actions of the command's staff.

3-60. PA SOPs differ from PA plans and PA annexes to OPLANS in that they specifically detail and describe how PA is conducted within a certain command or unit. They are routine procedures and actions that apply to each section or unit.

3-61. The senior PA NCO prepares the staff section or unit PA SOP. PA units designated to support or augment specific commands in the execution of contingency missions should use SOPs from these supported commands.

3-62. SOPs should address:

- Preparation for combat. Stockage, prepackaging, and maintenance of vehicles, equipment, and expendable and nonexpendable supplies.

- Vehicle load plans.
- Alert and mobilization actions, routines and procedures.
- Composition of quartering and/or advance parties and rear echelons.
- Organization for combat, including detailed delineation of duties for each individual, shift compositions, and plans for reconstitution in the event of combat losses.
- Operations center and media center layouts (theater, corps and/or division main/rear CPs).
- Procedures for preparing, disseminating and disposing of records, reports, estimates and orders.
- Physical, document, and tactical security.
- Communications procedures. These steps include radio/telephone operating procedures unique to the command, message routing and preparation formats, and operation of communications and data transmission equipment.
- Movement and displacement.
- Operations under NBC conditions.
- Field Maintenance.
- Personal hygiene, rest, and morale, welfare and recreation requirements and procedures during deployment.
- Post-operations and reconstitution procedures. Maintenance, restocking and packaging composition of advance and rear parties; disposition of records, and preparation of after-action reports are included.
- A PA SOP outline is included in this manual at Appendix K.

POST-MISSION PLANNING CONSIDERATIONS

3-63. During mission planning and preparation, Public Affairs planners should consider debriefing and other post-mission activities. These activities normally include:

- Collective debriefing of the operational element on all aspects of mission execution, including lessons learned.
- Collection of maps, notebooks, logbooks, plans, annexes, duty officer/NCO logs, serious incident reports, news releases, tapes and transcripts of news briefings and conferences, and all other information products pertinent to the mission after action report.
- Maintenance and storage of unit and personal equipment.
- Individual debriefing of key personnel.
- Other reconstitution measures as required.

3-64. Upon completion of these activities, the operational element begins pre-mission sustainment training or prepares for its next mission. The planning staff begins review of lessons learned for integration into future plans. See Appendix W for information on producing PA Lessons Learned for the Center for Army Lessons Learned.

Chapter 4

Media Facilitation

"It is likely that small pools of news media will be assigned directly to operational units to cover all facets of activity. With few exceptions, there will be no security review of media copy or audiovisual products. The policy will be to maintain security at the source. It is important to support the efforts of the media and our dealings with them should not be confrontational, but professional and courteous."

- GEN Binford Peay
Commander, U.S. Central Command [1994]
FM 3-61 (46-1), Public Affairs Operations

PRINCIPLES OF INFORMATION

4-1. In the past 20 years the Army has undergone a fundamental shift in our approach to dealing with the news media. In response to the perceived treatment by the press during the Vietnam War, we have gone from adopting an exclusionary tactic for the conduct of the invasion of Grenada in 1983, to managing the controversial pool system for covering the initial stages of the Gulf War in 1991, and more recently evolving into an almost completely open access policy in Somalia (1993), Haiti (1994) and Bosnia (1995) operations.

4-2. The results of this policy evolution are the "DoD Principles of Information," which form the foundation for the PA function of media facilitation. The basic approach that DoD and the Army take to media facilitation is contained in Appendix A, The DoD Principles of Information and Appendix B, The Guidelines for Coverage of DoD Combat Operations.

4-3. Simply stated, media facilitation is providing assistance to civilian and military news media representatives covering an operation. The objective of media facilitation is to support news media efforts. This includes providing accurate, timely, balanced, credible coverage of the force and the operation, while minimizing the possibility that media activities will disrupt the operation.

4-4. Media facilitation includes assisting media entry into the area of operations, registering media representatives, orienting them on the ground rules for coverage and ensuring that they understand the security policies and constraints under which they must agree to operate if they desire Army support.

4-5. Media facilitation also involves arranging interviews and briefings, coordinating unit visits and escorts, and assisting media representatives with transportation, messing, billeting, communication support, safety and equipment. Media facilitation involves the early establishment of a media center as a focal point for media wishing to cover an operation, for Army personnel seeking assistance with media representatives in their area, and for resolution of problems or incidents resulting from media-military interaction.

4-6. A primary strategic goal of any Public Affairs staff is to support an operational commander in achieving a constant flow of complete, accurate and timely information about the mission and U.S. forces.

4-7. The PA staff accomplishes this goal by making information fully and readily available within the constraints of national security and OPSEC, and by facilitating inclusion of civilian and military news media representatives in military units whenever possible.

PA STAFF AND MEDIA CENTER RESPONSIBILITIES

4-8. **PA Staff Sections**. The prime focus of the PA staff is staff coordination. The staff will be the element tasked with executing the media facilitation strategy.

4-9. The staff provides PA planning and operational guidance to the PAO. They ensure leaders within the command understand the commander's media relations policies, and serves as the command ombudsman in the settlement of conflicts between the media and the military.

4-10. As an active participant in the command's information planning element, the staff coordinates with G2, G3, G5, PSYOP, U.S. Information Service, and other staff elements in developing the commander's information strategy to ensure synergy and to reduce the probability of conflicting messages.

4-11. **Media Operations Centers**. Currently, joint, combined and Army media centers fulfill the requirement for a focal point for the news media during military operations. In essence, the media operations center (MOC) is a command post for media support efforts. It serves as both the primary information source, and as a logistical support and coordination base for commercial news organizations covering the operation.

4-12. Media centers are organized when large numbers of news media representatives are anticipated to cover military activities. Media centers may be formed for all types of operations or for any stage within an operation.

4-13. When operated by unified/specified commands, these media centers may be called a Joint Information Bureau (JIB). At the combined commands, they are called an Allied Press Information Center (APIC), Coalition Press Information Centre or Combined Information Bureau (CIB). At theater level and below, they are simply referred to as Media Operations Centers.

4-14. MOCs support the commander and are subordinate to the command's PAO. They provide the commander a professional, immediately available, fully trained organization designed to respond to national and international civilian media interest in American military operations.

4-15. In addition, the media operations center provides the following functions:

- A single point of contact and information source for media within the theater
- Briefings and enforcement of media guidelines and ground rules
- Primary information release authority for the senior PAO
- Coordination of news media coverage with corps, divisions, brigades, etc.
- Coordination with all service branches for each service, agency or country
- Identification and communication of host-nation sensitivities to all personnel in theater
- Preparation for and conducting press briefings and news conferences
- Registration of news media personnel
- Media Operations Center Staffing and Organization

4-16. Organization and personnel staffing of media operations centers are determined by the responsible command in coordination with the PAO and his staff. APIC staffs should be a proportionate representation of the forces, with representation from all services involved in the operation. This will be determined by CPA at the unified command. Regardless of the echelon establishing a media operations center, the organizational model is functionally designed and remains relatively the same.

4-17. MOCs normally consist of two major elements: a Headquarters Group and a Media Operations Center Group.

4-18. The headquarters is made up of the command group and support staff. The command group contains the commander, deputy commander and/or executive officer, and the sergeant major. The support staff is normally task organized to support tailored forward deployed MOC teams or sub-MOCs when the APIC operates as other than a single element. The support sections provide administrative support, conduct lease and purchase contracting, setup, operate and maintain the unit's equipment, and conduct the day-to-day operation of the MOC. The support sections are responsible for the execution of MOC communications, supply operations, administration support, vehicle maintenance, security and other support functions as required.

4-19. The Media Operations Center consists of a Plans Section, a Media Support Section and an Information Operations Section.

4-20. The plans section is responsible for all MOC media planning. It establishes MOC requirements and determines operating procedures and policies. It maintains channels of communication with OASD(PA) and the JPAO (or senior command PAO). It is responsible for recommending and assisting in the development and dissemination of PA Guidance. It monitors available major U.S., international and local television and radio broadcasts and print publications providing coverage of the operations, conducts news media analysis and evaluates the effectiveness of MOC operations.

4-21. The Media Support Section (MSS) is the primary point of contact for news media representatives (NMRs) in an area of operation seeking information or assistance in covering the force and the operation. The MSS receives and registers NMRs, briefs NMRs on the media ground rules and security procedures or concerns, and orients them on the force, the operation and other pertinent issues (special safety or host nation considerations).

4-22. The MSS orchestrates the command's news briefings and coordinate for subject matter experts to explain and discuss operations and capabilities. The MSS is also responsible for coordinating for appropriate, knowledgeable escorts, unit visits, and service member interviews. It assists the Joint Force or other senior PAO in preparing service members for interaction with the news media. Finally, it provides support to Joint Force elements and service component PA elements seeking assistance with NMRs.

4-23. The Information Operations Section is responsible for monitoring plans and operations from within the command's operation center and assessing the PA implications of events occurring throughout the area. It ensures that the MOC has current situation information, is aware of issues of potential media interest, and can obtain any operational information necessary for the development of responses to media inquiries in a timely manner.

4-24. The IO section ensures that PA operations are synchronized with other combat functions and promote early coordination of PA, CA and PSYOP functions.

4-25. **MOC Staffing**. Currently, media operations center staffing requires augmentation, either by PA-trained individual fillers or by Army PA units. As fully independent units, the Public Affairs Operations Center (PAOC) (SRC 45423A000) and Mobile PA Detachment (MPAD) (SRC 45413A000) are currently organized, trained, and prepared to fill this role.

4-26. In fact, these Army PA detachments are specifically designed to function as an Army media operation center in theater, corps, or division-controlled operations. MPADs can be combined to form media sub-centers in forward battle areas.

4-27. PA personnel from non-deployed commands and installation PA sections may be called upon to augment news media centers however, requests for individual augmentation should be coordinated through operational channels. Reserve and Guard unit personnel can be used to augment on a voluntary basis.

4-28. An example of a media operations center is included at Appendix L.

MOC OPERATIONS

4-29. In major operations -- actions conducted by unified commands -- a Joint Information Bureau will usually be the first to deploy for this purpose. A JIB will be staffed by public affairs personnel from the services represented in the joint force; participating services may establish their own media centers subordinate to the JIB to disseminate information about their particular missions.

4-30. As the operation unfolds, the Army plans for and contributes to a replacement PA organization for the JIB which consists of individual PA personnel from each of the services and Army PA detachments.

4-31. Media Center operations will be based on five primary assumptions:

- Accurate information is available in a timely manner and adheres to the DoD Principles of Information in Appendix A.
- Current trends in communications technologies within the information environment will continue to reduce the news media's reliance on military support and assistance when covering operations and will continue to increase the availability of information to a worldwide audience.
- Media representatives will be in an area of operations at the start of, and in most cases, before an operation begins.
- Media interest and coverage in non combat operations may be higher at the outset, and barring a significant event which renews national or international attention or interest, will taper off over time. During a high-intensity conflict, media interest could remain high.
- Military PA elements require access to complete information, state-of-the-art communication equipment, and must possess sophisticated coordination channels in order to pre-empt speculative, inaccurate or biased reporting.

4-32. Media centers will support and be responsible to the senior commander of the operation on a 24-hour basis. Media centers are usually established by unified command CINCs to support the news media in an area of operation.

4-33. During the first 24 hours after arrival in a new theater of operation, a media center can provide limited media support services.

4-34. Within this first operational day, the MOC must:

- Establish a "hasty media center" as the initial focal point for the news media until additional media support forces arrive.
- Establish communication with OASD (PA), each service's PA chain of command, and with units operating within the theater.
- Request operational information release authority within the theater.
- Establish command structure/lines of authority.
- Coordinate with appropriate authority for leasing and purchasing contracts.
- Begin to register news media personnel in the area
- Provide basic media support (coordination of media access to subordinate units and media escort as resources permit).
- Assist or conduct command news briefings and conferences.
- Coordinate Subject Matter Expert (SME) interviews.
- Be capable of assisting in the transmission of media products.

4-35. Media Support -- Initially, the media operations center will need to provide varying degrees of support to news media personnel including specialized equipment (flak vest, NBC gear, helmets), transmission of media products, etc.

4-36. This support may include but not be limited to:

- Coordinate media contact with units or individuals to include SME interviews
- Provide a single point of contact for information on operational issues
- Provide news releases, fact sheets, copies of transcripts for news briefings/conferences and copies of archival file products
- When other means are not available, the media center may provide coordination for transportation (to and from interview sources), transmission of media products and food and billeting
- Provide limited media escort within the area
- (SOPs for MOCs should be pre-established for each theater of operation and used for media operations within that theater.)

REGISTRATION OF MEDIA

4-37. Principle to supporting the commander's information strategy is the inclusion of news media representatives (NMR) within Army units from the earliest pre-deployment stages of all operations. The personal safety of media representatives, as acknowledged by the media themselves, is not a reason for excluding them from operations.

4-38. However, all media requesting support or access to units to cover Army operations must be registered. This includes freelance journalist, military media representatives, such as those who are assigned to Armed Forces Radio and Television Service, Stars & Stripes newspaper and other Armed Forces Information Service (AFIS) personnel who are not supporting units on the battlefield.

4-39. **Registration versus Accreditation**. Accreditation is the verification and validation that a person represents a legitimate commercial news organization. This means that accrediting governments or military organizations will physically verify the affiliation of an applicant with a specific news organization.

4-40. This is difficult to perform amidst an ongoing operation, especially when deployed far away from CONUS. It is generally accepted that, when overseas, the decision to accredit news media is made by the host nation's government in coordination with the combined or unified commander.

4-41. When accreditation isn't required by the host nation, responsibility for this determination is held by the combined or unified commander. Accreditation is normally performed at Corps level or higher.

4-42. Accreditation is a major problem for many commands because they are forced to determine the legitimacy of smaller, lessor-known news organizations and freelance journalists without news organization affiliation.

4-43. Unless it is absolutely required by host nations, the American military will attempt to avoid accreditation.

4-44. Registration, however, is merely an accounting tool, which provides PAOs the ability to know what media are represented in the theater, where they are located, and their movement around the theater. This information is helpful in planning and conducting media logistical support and transportation, and in preparing subordinate commands for media encounters. It is also helpful to commanders who might want to provide newsworthy events to the media.

4-45. Registration also identifies which news media have asked for military assistance and access, and have agreed to the command's media ground rules.

4-46. **Registration Requirements**. The registration process is conducted in five basic steps:

- Verify the identity of the media representative (including checking for valid passport/visa, professional media organization membership card, media ID card, other military press credentials, etc.).
- Have them sign an agreement to abide by the established media ground rules for the operation in exchange for granting support, access to units, information and other privileges. If required, revoke credentials for those who violate the ground rules. (Enforcement of this requirement is essential.)
- Have NMR agree to and sign a liability waiver that frees the military of responsibility if the NMR is killed or injured as a result of covering the operation. (An example of a waiver of liability is at Annex I).
- Give NMRs proof of registration (memorandum, press badge or other identification).
- Maintain a roster of registered NMRs and monitor their movements during the time they are receiving military support.

4-47. NMRs who refuse to agree to the military ground rules and who are not registered will receive only the support and information assistance as provided to the general public.

4-48. NMRs should be informed that registration and acceptance of media ground rules will entitle them to better access to units and subject matter experts, and provision of military ground and air transportation when possible.

MEDIA GROUND RULES

4-49. Media ground rules will assist inprotecting the security and the safety of the troops involved while allowing you the greatest permissible freedom and access in covering the story. All interviews with news media representatives will be on the record.

4-50. Security at the source will be the policy. (An example of media ground rules is in Appendix X.)

4-51. The following categories of information are releasable:

- Arrival of major U.S. units when officially announced by a U.S. spokesperson. Mode of travel (sea or air) and date of departure from home station

- Approximate friendly force strength figures, after review by host nation government
- Approximate enemy casualty and POW figures for each action operation
- Non-sensitive, unclassified information regarding U.S. air, ground and sea operations (past and present).
- Friendly force size in an action or operation will be announced using general terms such as multi-battalion or naval task force
- Specific force/unit identification/designation may be released when it has become public knowledge and no longer warrants security protection
- Identification and location of military targets and objectives previously under attack
- Generic origin of air operations such as land or carrier based.
- Date/time/location of previous conventional military missions and actions as well as mission results
- Types of ordnance expended will be released in general terms rather than specific amounts
- Weather and climate conditions
- Allied participation by type of operation (ships, aircraft, ground units, etc)

4-52. Information Not Releasable

- Information about future military plans, activities or operations
- Vulnerabilities or weaknesses on command, control, personnel or the operation
- Friendly unit and command strengths, on-hand equipment or supplies; the presence, activities and methods of operation of specifically designated units or equipment
- Information on friendly force security and deception measures and countermeasures, and intelligence collection activities
- Specific information on friendly force current operations and movements, deployments and dispositions
- Information on in-progress operations against hostile targets
- Information on nuclear, biological or chemical weapons, equipment or training

MEDIA POOLS

4-53. Journalists, as a group, are strongly opposed to media pools in any form. The media pool is seen as a restriction placed on the media representatives and their ability to provide coverage of the news. They are grudgingly tolerated, and should be only used as a last resort when space onboard military transportation is limited, access to an area must be controlled, and after all other possibilities have been explored and eliminated. Even under conditions of open coverage, pools may be appropriate for specific events. Both the Army and the news media are in agreement, however, that limited access is better than no access at all.

4-54. When a pool system is required, the military PAO will identify the maximum size of the pool that can be supported. The news media representatives on the scene will select media pool members. A roster of media personnel registered with the Army PAO will be used to identify the media representatives eligible to participate. The pool should consist of, but not be limited to, a minimum of one video crew (camera operator, sound technician and reporter), one still photographer (wire service, newspaper, or magazine), one radio reporter, and one newspaper or wire service reporter. Special consideration must be given to international reporters as well. While this is a fair and representative pool structure, it is the media themselves who must determine the make-up of the pool. Some news events and situations may lend themselves more to print, or conversely television reporting, and the media representatives may choose to select an unbalanced pool.

4-55. All pool members must be willing and able to meet deadlines and supply information products (video, audio, still media, and text) in a timely manner to all media representatives who are entitled to material generated by the pool. The military media center will also have access to this information and will make it available to all other requesting news media organizations.

4-56. Consistent with its capabilities, the military will supply PAOs with facilities to enable timely, secure, compatible transmission of pool material and will make these facilities available whenever possible for filing independent coverage. In cases when government facilities are unavailable, journalists will, as always, file by any other means available. The military will not ban communications systems operated by news organizations, but electromagnetic operational security in battlefield situations may require restrictions on the use of such systems.

4-57. Once a media pool has been selected, the media pool will select a team leader. It is the responsibility of this team leader to ensure that members of the media pool meet their obligation to share information. The Army PAO will not involve himself in settling internal disputes of the media pool.

4-58. Finally, the pool is an option of last resort. It should be disbanded as soon as free and open access to the operational area can be allowed, normally within the first 24 hours of an operation.

THE DOD NATIONAL MEDIA POOL

4-59. The DoD National Media Pool was established to prevent recurrence of problems encountered with media coverage during Operation Urgent Fury in Grenada in 1983. During the first 24 hours of Urgent Fury, more than 600 reporters attempted to gain access to the operation. The large numbers overwhelmed the limited Public Affairs elements available to assist them.

4-60. In 1985, the Secretary of Defense established the DoD National Media Pool, a civilian news element of approximately 16 media representatives from various national news organizations, with the mission of covering an operation from its initial stages until open coverage could be allowed.

4-61. The pool members remain on call in Washington, D.C., and are available for immediate worldwide deployment. Their products are shared by the open news media until the pool is disbanded and access is granted to the entire news community.

4-62. Supported commanders are responsible for providing operational support to the DoD National Media Pool. At a minimum, the pool members will require:

- Daily, comprehensive and unclassified operational news briefings.
- Access to ongoing combat operations. The media are aware of the personal risks involved in covering combat operations. They will not be denied access to them based on risk to their personal safety.
- Reasonable access to key personnel. All information gathered from these personnel is unclassified and on the record.
- An escort -- usually a lieutenant colonel or colonel -- to coordinate pool support and access requirements.
- Transportation and itinerary planning and coordination that will allow media to gain access to the theater of operations and to disperse pool members throughout the operational area.
- In today's global information environment, when news media can report live from almost anywhere in the world in almost any environment, the technological capabilities of most news organizations decreases the importance of the DoD National Media Pool once word of an operation has spread.
- When the DoD media pool is operational, PAOs will attempt to provide the same information support concerning theater operations to all other media in the operational area.
- However, their primary responsibility is to the DoD Media Pool. After the DoD media pool is dissolved, all media in theater will be dealt with in an equitable manner with respect to information and support provided.
- As soon as open access to the operational area can be allowed (normally within the first 24 hours of an operation), the DoD National Media should be disbanded.

NEWS BRIEFINGS

4-63. There are several reasons for holding news briefings, in addition to the daily operational news briefing required at the unified command level.

- Credibility: The physical presence of a briefer and his willingness to meet the issue head on leads to a much more credible presentation
- Uniformity: All media get the same information at the same time.
- Expression of concern: A briefer represents the face of the command, which shows more concern than an impersonal news release, especially in situations where there is loss of life or extensive damage.
- Complexity of material: Where material is technical or complicated, the news briefing makes the subject matter more easily understandable. The question and answer session that accompanies a news briefing saves time in call-backs by news reporters needing clarification.

4-64. News briefings should be done daily during an operation and when important events dictate. They should:

- Get out a specific message
- Explain complex or technical matters
- Reach a large number of media interested in the same subject matter area.

4-65. PA specialists should think about media deadlines and set the time to help the media meet those deadlines. Be sure to invite all media within the area in a timely manner.

4-66. A knowledgeable and articulate spokesperson should be chosen to present the material. This should be the subject matter expert (SME), but may be the PAO or the commander. At the very minimum, a person of prominence within the command should be selected.

4-67. Other SMEs may be in attendance at the briefing to field technical questions. The SME interviews should be at the request of an individual media representative and the time should be set to facilitate the media to meet those deadlines.

4-68. Consideration should be given to the appearance of the presentation, the message, space, lighting, electrical needs, suitable setting, chairs, tables and press packets. All handouts should be reviewed.

4-69. Appendix H provides briefing and press conference formats.

Chapter 5

Information Strategies

The defining trend of the 1990s, from corporate boardrooms to private living rooms, is connecting everyone and everything to everyone and everything else. The American public is faced with many choices: interactive television, cellular phones, modems, faxes, personal digital assistants, an assortment of daily newspapers, online computer information services and other easily accessible information services -- if it can be connected, it seems to be in demand. The computer has invaded the American home. In this decade it has become routine that news and information can now be tailored to fit the individual needs of the consumer using on-line services, distributed electronically and received instantaneously at his or her computer. With the sophistication, power and miniaturization of these technologies improving each year, and the cost dropping at an equally rapid pace, the public will expect access to these devices and services and the information they carry. These emerging technologies have contributed to the refocusing of the Public Affairs mission. It has resulted in shifting the information provision function from an emphasis on producing specific products (such as post and field newspapers and radio/television news programs) to focusing on the processing of our themes and messages and their intended effects -- the function of information communication, rather than the form. This chapter explains the objectives of information strategies, identifies and explains the elements of information strategies, and describes the relative advantages and disadvantages of present day and emerging information communication channels available to PA organizations. Most importantly, it explains how best to use these information channels to satisfy the information needs of the various target audiences as we enter the information age.

INFORMATION STRATEGIES

5-1. Information Strategies is the sum of all actions and activities, which contribute to informing the American public and the Army. The responsibility for this activity is assigned to an element within each PA section, which focuses entirely on accomplishing the information strategy mission. This section is usually called the Public Affairs Information

Services Section. At all echelons, it employs numerous techniques to provide news and information to internal and external audiences. The Army provides an expedited flow of complete, accurate and timely information, which not only communicates the Army perspective, but also attempts to educate audiences and engender support for the force.

5-2. Using a combination of contracted services, organic military assets, and government and commercial communications networks, Public Affairs organizations provide information to news media representatives, deployed soldiers, home station audiences and the American public. The Information Services Section within a Public Affairs organization coordinates information efforts and develops informational products (such as digital text, graphics, and photos, printed publications, audio/video news releases and graphic imagery) into consolidated campaigns designed specifically to present the Army's perspective. This means that Army Public Affairs communicates information to create an informed American public and Army force, assist them in gaining a clear understanding of the strategic, operational and tactical situation.

5-3. To accomplish this, the Information Services Section must develop information objectives or goals during the planning process prior to an operation. These information goals are similar to the PA standards of service and support, which appear in Appendix Q, in that they establish a basis for determining successful information communication operations.

5-4. These information objectives should include:

- Ensuring an understanding of the role of America's Armed Forces in American society.
- Ensuring an accurate perception of the particular military situation or mission.
- Ensuring an understanding of individual and unit roles in mission accomplishment.
- Establishing confidence in America's Army to accomplish the assigned mission in accordance with our national values.
- Establishing confidence in and support for American soldiers.

5-5. By establishing a comprehensive information strategy program, Public Affairs can assist in mission accomplishment by increasing audience understanding of the situation and establish confidence in and support for the force. This contributes to unit cohesion and provides commanders with increased range of action, free of distractions and limitations.

5-6. This is best accomplished by three basic types of information campaigns:

- **MISSION.** Both external and internal publics need to know what the mission is, what they're being asked to do and why. They need to know not only the organization's mission, but also how it fits into the big picture -- the political/strategic-level situation, and why it is important.
- **ROLE.** All military members and civilian employees need to have an understanding of their job and how it relates to mission accomplishment. The general public needs to have an accurate understanding of the military's role and its ability to accomplish

the mission. This understanding results in confidence in the force and demonstrates American unity and resolve.

- **MORALE.** Military members need to have access to news and information about current events and the activities available to them while deployed. They also need to have access to information from civilian commercial news sources. This is important because, in addition to being more credible, it allows the deployed force to see how the operation and their participation in it are being portrayed for the American public. In order to better understand the mission, their role in it, and give it his or her full effort, they have to know what effect the operation is likely to have at the local, regional, national and international levels. The opportunity to involve themselves in educational and other activities is necessary to quality of life and morale. A well-informed service member is more effective.

5-7. The general public is interested in soldiers, their lifestyle, how they are being treated and their ability to accomplish a given mission. Information about these topics provide reassurance, confirming that soldiers maintain professional and ethical values and are being cared for adequately.

Historical Perspective

During Operation Desert Shield/Storm, many commanders developed innovative methods of sharing Command Information products produced in theater and in the rear. The products greatly enhanced morale at both ends. Some commanders and PAOs used the products as issue management tools to dispel rumors in theater and at home station. The products included field newspapers, newsletters, videotapes, audiotapes, etc. They also let soldiers returning from the area meet with family support groups to answer any questions.

(After Action Report 1991)

RESPONSIBILITIES

5-8. Commanders at each echelon are responsible for Public Affairs operations and support. Public Affairs officers and noncommissioned officers at various levels assist commanders in the discharge of these responsibilities. Public Affairs staffs are responsible for accomplishing the Public Affairs information communication mission. This responsibility includes Public Affairs operations in all subordinate, assigned or attached commands.

5-9. PA is only one of many information channels available to a commander. PA information provision cannot substitute for a commander's personal involvement in his "Command Information" program.

5-10. All public affairs practitioners have access to all information that is not classified or violates Operational Security or the Essential Elements of Friendly Information (EEFI) for use in preparing information products. Commanders must ensure the EFFI is up-to-date based on current situations and operational guidance. Public Affairs personnel at all levels must produce and release accurate information packages based on DoD directives and Army policies.

5-11. **Strategic Level Commands** are responsible for providing public affairs guidance to subordinate units. They develop central themes and messages and provide umbrella guidance to subordinate PA staffs. They must also provide subordinate commands with information useful in preparing information products for internal and external release. They are additionally responsible for marketing public affairs information products to subordinate commands, home stations, the Army as a whole, as well as the general public. Strategic level Public Affairs staffs are the primary coordination point for the Armed Forces Radio and Television Service and the geographical manager for radio and television services including personnel, down links, facilities and equipment.

5-12. **Operational Level Commands** are responsible for communicating public affairs guidance to subordinate units. They expand on information campaign themes and messages, and provide additional information products to subordinate command PA staffs. Additionally, they provide subordinate commands with information useful in preparing information products for internal and external release. They are responsible for gathering and producing public affairs information products for release. In the event the Operational Command is the senior command in an area or theater, it assumes the responsibilities of the strategic level command.

5-13. **Tactical Level Commands** are responsible for gathering information products for release through their next higher headquarters to home stations, the Army as a whole, as well as the general public. These commands are also responsible for coordinating the dissemination of information and information products received from senior commands down to subordinate commands. In the event the tactical command is the senior Army command in an area or theater, it assumes the responsibilities of the operational-level command and will be augmented to accomplish these additional functions.

INFORMATION STRATEGY PROCESS

5-14. The Information Services Section uses all available means to gather complete, factual, unbiased information for use in information campaign development. The information is developed, converted into the most appropriate product form based on the information needs/target audience assessment and information communication channel availability, and then transmitted to the intended audiences. This is called the Information Provision Process.

5-15. Although the information strategy process follows a deliberate cycle, it is a continual process. Information campaigns are also conducted simultaneously, with personnel examining external and internal information needs, carrying messages from concept through execution and program evaluation, to accomplish specific PA objectives. The cycle has four phases -- acquire, process, protect and distribute. (See Figure 5-1.) Evaulation is a key component in the cycle. It must be conducted throughout the four phases. This ensures the campaigns are meeting their objectives, and are altered if they do not.

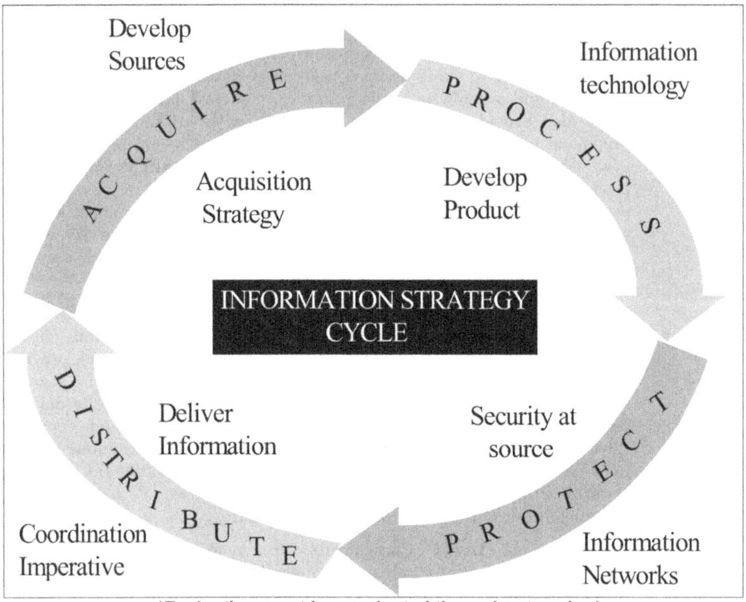

*Evaluation must be conducted throughout each phase

Figure 5-1

ACQUIRE

5-16. During this phase, a ISS identifies and assesses several factors: the situation, the environmental factors, the mission, and the target audiences. They determine the information needs of the various target audiences. They then begin to gather information based on the information requirements of these audiences.

- ***Information Needs Assessment and Audience Analysis.*** Identification of the target audiences and target audience populations and densities must be developed during the planning stages of operations with target audience assessments. Upon assessment of target audiences and consideration of which type of information product will best serve each audience, commanders must ensure adequate public affairs personnel, equipment (to include communications), resources and funds are available and included in OPLANS/OPORDS to achieve mission success. An assessment of soldier information needs is crucial to the information package development and the selection of appropriate products.
- ***Information Gathering.*** Information gathering is the first step in producing packages for release to internal and external audiences. Public affairs personnel gather information through

operational, administrative, logistical, battle, staff, command and support channels. They also acquire information through the media, research, leaders and the culture at large. Information may be obtained via electronic and telecommunication systems in addition to written documents and oral communications. While information often comes from superior and subordinate commands, it may also be obtained laterally.

- Sources of information must be valid and diverse enough to provide a broad overview. Public Affairs staffs must ensure the source does not speculate, nor speak out of his area of expertise. However, soldiers' experiences and personal opinions may lend credibility and provide a "grass-roots" view.

PROCESS

5-17. During this phase, the ISS begins to develop products, prepare them for release and determine methods for distrubition.

- **Product Development.** Development of information products is performed to some degree by PA personnel at all Public Affairs levels and is an on-going process. The term "product," as used in this process, means the message for intended communication, regardless of the format or communication channel proposed. Initial production development may be command directed or initiated by the Public Affairs staff or provided by other Army agencies. The command's resources and the target audience's requirements will determine the product type.
- **Prepare products**. Information packages should be prepared in accordance with Public Affairs Guidance, Soldiers' Manuals, Field Manuals, SOPs and Army Regulations. Ultimately, the Public Affairs staff is responsible for content.
- **Media Forms/Methods.** Soldiers use a variety of technologies to gather information and produce information products. During the gathering process, PA soldiers conduct interviews, attend briefings, withdraw data from government and commercial computer databases, bulletin boards, and e-mail systems. They acquire text, graphics, photography and motion video from government and commercial Internet systems. The nature, distribution, usability and flexibility of public affairs systems are crucial in the processing of information.
- Professional quality systems should be used whenever possible. For printed products, preferred systems include computers, desktop publishing, word processing, laser printing, etc. Reproduction may be Army-contracted, Army-funded or reproduced using the command's assets. Video and audio products intended for release to news media should attempt to be broadcast-industry standard.
- Electronic newsgathering and editing systems should be used when available. Visual products should be generated by modern methods including digital imagery and computer graphics.
- **Print.** Articles released to home station for military publications and family support group publications; for marketing to civilian publications; field publications, e.g. newsletters, with and without photography.

- **Video.** Raw electronic news gathering video and printed news scripts for release to military and civilian outlets.
- **Audio.** Radio interviews; features; internal command information scripts; radio news; news reports for release to military and civilian outlets.
- **Visual.** Digital imagery; photographs; slides, view graphs; graphics for release to military and civilian print and broadcast media.
- **Digital.** Each of the categories described above may be developed and distributed electronically, either through commercial information services directly on the Internet, or by using tactical Army communications systems (SINCGARS and ATCCS). Modern technology in use on the battlefield has made digital transmission the preferred method for all types of products.

PROTECT

5-18. **Security at the source.** No information strategy is complete without a clear cut understanding of how to protect the information. Both sides can benefit from information and use information simultaneously against each other. Pieces of the right information can have a dramatic impact on the outcome of an operation. Public Affairs professionals will continue to protect vital information by practicing security at the source and following established operational security guidelines.

5-19. In addition to protecting the raw and completed information products, public affairs personnel must also take the necessary steps to protect information networks.

DISTRIBUTE

5-20. **Dissemination.** Public Affairs information packages should be released in the format most easily used by the recipient. While this is not always possible, a product stands less chance of being used if it is incompatible with the recipients' equipment. For example, a video product released on Hi-8 to a TV station that works exclusively with Beta SP has less a chance of being aired than a video story in a compatible format.

5-21. Public Affairs information packages must be expedited to the users by the most technologically advanced reliable method. Great consideration must be given to the speed and reliability of the mode of dissemination. This must be included in target audience assessments and conducted during planning stages of operations. Articles and photos may be sent from deployed locations to home stations via computer systems and telephone lines. Yet based on the quality of the telecommunications system, it may be more reliable to use the mail or a courier. As technology improves PA capabilities, Public Affairs will incorporate those improvements into the information gathering and dissemination system to increase its potential to reach an ever-growing, information-hungry public. For example, the emergence of smaller, more powerful satellite link ups can provide PA elements the ability to reach targeted audiences sooner and from more locations.

5-22. Internal information packages must be available to soldiers at all levels of command. Public Affairs must develop and coordinate a distribution scheme with the commander, the general staff, and with signal as the proponent for physical distribution of certain commercial news and information products. An efficient distribution system will also ensure prompt delivery of public affairs products. The public affairs staff must conduct periodic quality control checks and update the distribution scheme as necessary, based on changing population densities or information products. Electronic means are the preferred mode of distribution, however additional methods include, but are not limited to contract delivery, AG distribution and the military postal system.

5-23. **Products for distribution to deployed soldiers.** Publications produced by other military agencies intended for deployed soldiers must be given the same distribution considerations as commercial information packages. For example Stars & Stripes, Soldiers Magazine, Army Trainer and home station post newspapers contracted for delivery to deployed soldiers must be given the same distribution considerations as other publications. However, a separate distribution scheme may be required.

INFORMATION PROGRAM EVALUATION

5-24. The final step in the Information Provision function is the evaluation of our communication efforts. Evaluating communications programs is research, which boils down to a series of questions:

- Did we achieve our objectives?
- Were our policies and programs effectively communicated?
- Was the operation affected positively by our efforts?
- Did the American public support our soldiers? Was unit cohesion enhanced?
- What audiences received our messages? What was the impact of our communication programs on these audiences?
- Research is the foundation of the Information Program Evaluation. Corrections and changes in courses of action should be based on solid factual information. Methods for conducting Information Program research are discussed in greater detail in Appendix P.

NON-HIERARCHICAL AND INTERNET INFRASTRUCTURE AS A RESOURCE

5-25. Modern technology has provided us with an advanced form of communications structure called non-hierarchical structure. The advantage of non-hierarchical communication structuring is that every "node" in a communications web shares information with every other "node". Each node on a network can identify itself and "find" others in the network in order to communicate specific information. Public Affairs elements act as information nodes, gathering, developing and sharing information vertically and horizontally on the battlefield.

5-26. While this technology was originally intended for command and control, it is essential for other functions on the battlefield as well. Well-coordinated public affairs operations will leverage this capability to move information -- sending messages around the battlefield, to and from home station, and up and down the chain of command.

5-27. The end state of this technology effort is that both organizations and individual soldiers on the battlefield will possess this capability.

TELECOMMUNICATION SYSTEMS

5-28. Information technology provides the means to collect, process, display, develop and disseminate information in an unparalleled manner. This technology has begun to revolutionize our approach to information provision.

5-29. The PA specialist now has the ability to access desired information on a certain issue and tailor and develop this information into a message for dissemination -- all from a personal laptop computer.

5-30. This "from anywhere to anywhere" capability allows the PA specialist at all echelons to accomplish his mission of presenting the Army's perspective -- framing issues and informing targeted audiences.

5-31. There are two telecommunication systems available to Army PA specialists: DoD's internal secure communication network, known as the Defense Data Network (DDN) and the worldwide commercial information network -- Internet. PA specialists must be familiar with both.

DEFENSE DATA NETWORK

5-32. Defense link is an entry point to Internet sites operated by the Office of the Secretary of Defense, the armed services and related defense agencies. It provides a means to search and download Department of Defense directives, obtain transcripts of important speeches and connect to other sites dealing with specific military operations.

5-33. The Department of Defense organizations are extensively represented on the Defense Data Network. Each of the military services has a centralized directory of its Internet sites. Army Link, as its name suggests is a catalog or hotlist of more than 280 Army home pages. The sites are indexed alphabetically and by subject area.

INTERNET

5-34. The Internet is a network of networks. Originally, the Internet was developed by DoD with the goal of building a automation system linking government agencies together which would continue to function in case part of the network ceased to exist. From this inception, the Internet grew, as other agencies and activities noticed the benefit of linked telecommunications. The system now includes organizations from all segments of society: government, defense, business, education, health services, etc. There are more than 30,000 networks participating on the Internet and more than 150 million individual users. And more are connecting everyday.

5-35. Army Public Affairs personnel have access to the Internet through their local Directorate of Information Management (DOIM).

5-36. On the Internet, there are six main activities: "Hypertext" documents with multimedia links on the World Wide Web can be viewed, files can be downloaded with a file transfer program or "FTP," data can be located, communication via e-mail, reading and posting messages to news groups and bulletin boards, and logging in to remote computers using Telnet.

5-37. You can do most of the same things in the Internet that you can do on any of the commercial information services, but you do them in the context of a much larger network that isn't centrally organized or controlled.

5-38. **World Wide Web.** The World Wide Web, also known as WWW, W3, or simply the Web, is one of the most popular activities on the Internet. The Web allows you to view documents that feature graphics, "hypertext" (text which contains highlighted keywords which are linked to other documents or information sources available on Web) and multimedia links. The hypertext and multimedia links are tied to other documents or information forms that might be on the same computer, one across the country, or a machine on the other side of the world.

5-39. The WWW contains thousands of main menus, called "home pages" which identify the various sub-categories at each web site. Web users can search using key words or locations through the lists of home pages for specific web sites.

5-40. **Gophers.** A gopher is a menu driven system that offers text only. Users of the web can access gopher sites and retrieve information. It is a predecessor of the Web.

5-41. **News groups.** They are discussion groups built around a particular topic. Some are managed, others are not. News Groups take two forms: Mailing List: A list of E-mail addresses to which messages are sent. You can subscribe to a mailing lists typically by sending an email to the contact address with the following in the body of the message: the word subscribe, the name of the list, and your email address. Discussion groups: A particular section within the USENET system typically, though not always, dedicated to a particular subject of interest. Also known as a newsgroup. USENET is a collection of the thousands of bulletin boards residing on the Internet. Each bulletin board contains discussion groups, or newsgroups, dedicated to a myriad of topics. Messages are posted and responded to by readers either as public or private E-mails.

5-42. **E-mail.** E-mail is a means of interpersonal communication that falls somewhere between the immediacy of a phone conversation and the more thoughtful but slower exchange of ideas previously done by writing letters and memos. The specifics of using it vary greatly according to the mail software being used.

5-43. **Commercial On-line Information Services.** Users cannot directly dial up to the Internet using a modem, but must gain access through an internet service provider. Many offices, universities and large businesses provide access for their employees. For access at home, a user can either subscribe to one of the commercial on-line services, which provide Internet access in addition to their other features or to one of the dedicated Internet server companies located in most cities.

Chapter 6

Training

"You must remember that whether you wear one stripe or six, one bar or silver eagles you automatically become an "Army spokesman" when you are approached by the press. Within 24 hours the words of that Army spokesman can be flashed world wide, particularly if they can be construed as criticism....... Everything you say should have the ultimate aim of furthering that effort. Your approach to the questions of the press should emphasize the positive aspects of your activities and avoid gratuitous criticism. Emphasize the feeling of achievement, the hopes for the future, instances of outstanding individual or unit performance and optimism in general. But don't destroy your personal credibility by gilding the lily. As songwriter Johnny Mercer put it, "You've got to accentuate the positive and eliminate the negative."

- U.S. Continental Army Command memo to advisers in Vietnam.
The Military and the Media,
William M. Hammond, 1988

TRAINING OBJECTIVES

6-1. The cost of learning in combat is too high. Therefore, commanders and staff section chiefs must conduct cost- and time-effective staff training during peacetime. In order to conduct effective training, every consideration regarding PA activities should be evaluated.

6-2. Evaluation should include, but is not limited to, analysis, planning and execution of the operation. The PA staff, in conjunction with other members of the organization, must train in the initial planning of the operation. In order to train effectively, the following considerations should be evaluated:

6-3. **Analysis.** Specified and implied tasks are identified at home station for deployment, mission and redeployment requirements.

- Feasibility of all tasks are determined and assessed.
- Consider any PA guidance issued by higher headquarters and determine specialized training requirements.
- Consider all phases of host nation operations and the effect on your operation and organization.

- Determine the availability of required assets and equipment for all phases of the operation.
- Consider the availability and use of internal and external assets -- joint, RC and combined -- to augment known shortfalls.
- Determine any transportation requirements for your organization or attached personnel.

6-4. **Planning**. Planning includes PA annexes, command directives, DOD guidance, regulations and directives stipulated by the commander's intent and mission requirements.

- PA annexes should contain a brief, general description of the situation and intent to conduct operations.
- Planning should also consider enemy forces and host nation personnel and their relationship to the operation.
- PA planning should be considered for all attachments and detachments currently known or listed under the task organization. This includes planning for assets and requirements from home station to redeployment.
- Training should be conducted emphasizing procedures for handling the transmission of all information. Theater-unique requirements may call for special training scenarios.
- All logistical and administrative requirements need to be addressed in the planning and training process.

6-5. **Execution.** The staff, soldiers and media understand the commander's intent in terms of his command position and operational mission. The operational area ground rules are included in the PAG and are widely disseminated throughout the organizations.

- Encourage local media to accompany deployed units.
- Adequate vehicles have been identified and assigned to transport PA personnel and media throughout operational areas. This includes vehicles, aircraft and available transportation assets for transporting all media personnel and products. Consideration of transportation should include all support activities in the forward area of operation.
- Public Affairs personnel in the media center will prepare releases of specific events to disseminate in the area of operations. News media inquiries are anticipated, received, researched and answered as quickly as possible.
- Local security review policies are clear and will not delay the transmission of information.
- Executions of the theater policy for registration policies and local media pool operations have been completed. Non credentialed media are identified and escorted to the rear for registration. Once registered, media personnel will be allowed to cover stories and interact with organizations in the area of operations.

6-6. The staff section must be able to:

- Cope with the unexpected. For example, media arriving at the unit level without PA escort or media not registered with the joint media operations center.

- Separate fact from fiction. For example, media having misinformation which they believe as fact and trying to confirm it.
- Coordinate well amid confusion. For example, be able to access the PA staff at the MOC.
- Think clearly while under great stress. For example, have a unit PA representative (UPAR) who will handle escorted/nonescorted media for the unit.
- PA elements develop training plans from assessments of their performance proficiency on their METL.
- Detailed training plans for Public Affairs Detachments are contained in ARTEP 45-500-10-MTP, for the Mobile Public Affairs Detachment in ARTEP 45-413-30-MTP and for the PAOC, ARTEP 45-326-50-MTP.
- The training plans in those documents can be readily adapted to train other types of PA elements, such as a division PA staff.

PRINCIPLES

6-7. All PA training must be based on the training principles in the FM 25-series:

- **Train as combined arms and services team**. Do not train in a vacuum. Train with the unit you support.
- **Train as you fight**. Conduct realistic training. PA leaders must move soldiers out of the office and into the field to conduct training, including joint and combined operations. Set up a media operations center and conduct operations for a day.
- **Use appropriate doctrine**. All PA leaders should be familiar with FM 3-0 (100-5), *Operations*; FM 3-13 (100-6), *Information Operations*; and the PA manual, FM 3-61 (46-1), *PA Operations*.
- **Use performance-oriented training**. Performing tasks under field conditions with appropriate evaluation enables your soldiers to perform better under actual conditions.
- **Train to sustain proficiency**. PA units must take advantage of all training opportunities to sustain proficiency, participating in all CPXs, FTXs and deployments that their supported units are involved in.
- **Train to challenge**. Training for PA tasks must create the kind of pressure PA soldiers will face in actual situations. Ensure it is challenging but realistic.
- **Train using multi-echelon techniques.** The entire PA chain-of-command must participate in training for it to be realistic and effective.
- **Train to maintain**. Soldiers and leaders must keep equipment in a high state of readiness in support of training and for deployment.
- **Make commanders the primary trainers**. The unit commander is the primary training manager for the unit. The commander assigns primary responsibility to officers for collective training and to NCOs for soldier training.

TRAINING OPPORTUNITIES

6-8. Regardless of type, size or configuration, PA elements/units must train with the units they support. PA leaders must evaluate the ability of their units to support their real-world missions successfully and determine areas of weakness.

6-9. PA leaders must determine what training opportunities they can participate in realistically and coordinate for support from higher headquarters. If support is not provided, then unit leaders must create their own unit training program that simulates the tasks they will be asked to do when deployed.

6-10. For example, PADs should deploy with brigades to NTC or JRTC to practice their unit mission and capabilities and be evaluated accordingly. If the PAD is tasked with garrison duties, these duties must be retasked to allow the unit to train and maintain.

6-11. PA elements and units should take advantage of the excellent training opportunities offered by the JRTC and NTC. Both centers conduct media-on-the-battlefield training for visiting maneuver units. Public affairs elements should accompany the maneuver brigades and establish field media centers to take maximum advantage of these resources. It is imperative that PA elements be able to function effectively in austere environments. They cannot perform to their maximum capability unless they train accordingly. The key is repeated, tough, realistic training.

Historical Perspective

An initial PA team was sent to Somalia to set up a media operations center in December 1992. Mogadishu at the time was a war zone with many fronts. Initially, U.S. forces set up operations at the airport. Under the constant roar of C-5s, the PA team tried to organize a MOC. They had a very difficult time dealing with the lack of power and water. Eventually they were able to set up in a schoolyard. A "JIB-in-a-box" arrived with computer equipment, but it was all 220-volt — power in Somalia is 110-volt.

For the duration of the operation, personnel at the MOC learned to live and work under austere conditions, where food, water, power, and communications were constant problems. They eventually were able to set up a briefing room with benches, an old couch and a bulletin board for posting media opportunities, and conducted media operations under austere conditions.

(After Action Report 1992)

6-12. Home station CPXs and FTXs are fundamental training events that provide PA elements relatively low-cost opportunities to practice the full spectrum of wartime PA tasks on a reduced scale.

6-13. A visit by even a single reporter can be used to exercise a broad range of media support tasks including establishing a field media center; arranging interviews; preparing subject matter experts, individual soldiers and commanders to meet the media; preparing fact sheets; responding to media queries; registering media; arranging escorts and transportation, etc.

PA TRAINING FOR NON-PA AND PA PERSONNEL

6-14. Public affairs training is not limited to training for PA soldiers. PA elements/soldiers also have a responsibility to train non-PA soldiers and family members in military/media relations. Organic PA elements must develop training programs for their supported units. These can take the form of classes as well as pre-deployment briefings for units and family members.

6-15. Training for commanders and subject matter experts. Unit commanders and individual subject matter experts must be prepared to be interviewed. It is a PA responsibility to ensure they are familiar with the ground rules and know to restrict their statements and comments to their area of expertise. Details on how to do this are contained in the Soldier's Manual task, "Prepare a Spokesman to Address the Media," #224-176-4013.

6-16. Training for unit soldiers. Individual soldiers must be advised of the inevitability of media presence during military operations. Classes for units can be part of sergeant's time training and consist of teaching soldiers what they should and should not talk about when meeting the press. All soldiers should be encouraged to represent themselves as soldiers and encouraged to speak about the jobs they perform for the Army.

6-17. Training for family members. PA training for family members consists of educating them on their rights and responsibilities when interacting with the media. Family members often know more about specific operations than should be revealed to the media. They must be advised not to discuss information, which may be used by the enemy against their spouse's unit, such as details about troop movements, destinations, missions etc. They must also be advised that they have the right to refuse to talk to the media. Family member briefings should be a standard element of pre-deployment family support group activities.

6-18. Basic soldier skills. Public affairs soldiers are soldiers first and public affairs practitioners second. It is imperative they are well trained in basic soldiering skills. Public affairs elements and units must make time to train to standards on common soldier tasks that allow them to effectively shoot, move, communicate and survive on the battlefield.

6-19. Readiness. All PA elements and units must have, and exercise, detailed load plans. Soldiers should be aware of the importance of maintaining all equipment and vehicles in a state of readiness for deployment.

6-20. PA NCOs must ensure soldiers participate in preparations for overseas movement (POM) so they are administratively and medically prepared for worldwide deployment. Soldiers should be aware of the importance of wills, shots, powers-of-attorney, personal data and allotments.

6-21. Staff Section and Unit Training. A state of operational readiness to conduct combat or non-combat operations must be attained and maintained. This level of readiness is accomplished by preparing individuals, shifts and staff sections to perform assigned tasks and other duties at the desired standard of proficiency in advance of assigned operations.

6-22. Standards of performance must be set so the section can evaluate its performance. Examples of these standards include:

- Understanding DOD's policy statement -- Commanders will ensure maximum unrestricted disclosures of unclassified information to news media representatives consistent with operational security, guidance from higher headquarters and the privacy of individuals concerned.
- Escorted and unescorted media -- If the media has a PA escort, you may agree to an interview after the escort explains some basic ground rules. If not escorted, ask media personnel to accompany you to the command post, NCOIC or OIC and contact higher headquarters/public affairs officer.
- Conducting an interview -- Military personnel have the right to deny media interviews. If a soldier elects to provide the media with an interview, he should only discuss those things which he has direct responsibility or personal knowledge, and ensure an escort (PA or non-PA) is present. See Appendix N for more details.

THE INTERVIEW PROCESS

6-23. The Interview Process:

- Prepare for the interview. Consider the type of questions the media will ask, and think what your answer will be. When possible, ask for PA assistance (UPAR or PA personnel).
- Relax and be yourself. Imagine that the reporter is someone you know and talk with him in a relaxed manner.
- If cameras are present, ignore them and talk directly to the reporter. Be brief and concise. Remember, a TV news story will use only 10- to 15-second answers.
- If you need time to respond, ask the reporter to restate the question. A simple pause before answering the question is sufficient.
- Use simple language and avoid military jargon (i.e., military abbreviations or acronyms). If you must use military terms, explain what they mean.
- Use appropriate posture and gestures. Relax and be yourself.
- Answer only one question at a time. If asked multiple questions, answer the most important one first, or answer the one you're most comfortable with.
- Always try to end your comments on a positive point. This is your opportunity to tell your unit's story.

6-24. Things not to do:

- Do not allow media to videotape recognizable landmarks nearby, sensitive equipment, interior of tactical operations centers or other sensitive areas. Use OPSEC as guide on this matter.
- Do not answer speculative questions or give opinions concerning real or hypothetical ("what if") situations.
- Do not use the expression, "No comment." A more appropriate comment would be, "We don't comment on future operations." or

"I'm not qualified to respond to your question." or "That information is classified, so I can't discuss it."

- ***Never*** lie to the media.
- Consider everything you say to the media as "on the record." ***Never*** make "off the record" comments.
- Don't lose your temper when media representatives ask questions you consider inappropriate or foolish. Such questions are usually rooted in ignorance rather than in malice.
- Do not discuss operational capabilities, exact numbers or troop strengths, numbers/types of casualties, type of weapons systems or future plans. Use general terms like approximate, light, moderate or heavy.
- Don't repeat a negative phrase in response to a media representative's negative question (e.g., Q: Since your unit is poorly trained, can you really deploy? A: We're well trained and ready to go.).
- Staff section and unit training should be integrated into headquarters FTXs and CPXs in order to perform adequately within the unit's operational staff during real-world missions.

TRAINING EXERCISES

6-25. Training exercises will vary from major FTXs to CPXs and Tactical Exercises Without Troops (TEWTs). Each training opportunity provides the staff element realism, the opportunity to experiment and the ability to face situations. These challenging training exercises enable soldiers and units to tell the Army's success stories.

6-26. Training exercises also enable the commander and staff to:

- Emphasize the tactical SOP.
- Execute plans in a dynamic, hostile environment.
- Exercise bold solutions.
- Exercise contingency plans.
- Experience possible defeat without the penalty of combat loss.
- Work toward goal-oriented performance standards by team building while providing PA support.
- Training exercises also allow PA personnel to become operationally, strategically and culturally aware, and puts them in a unique position to interface with the news media personalities which shape the perceptions of our national and international audiences.

TRAINING FOR MEDIA FACILITATION

6-27. Facilitating media in the military environment includes three phases -- planning, preparation and execution.

6-28. **Planning.** The Media Operations Center receives requests for registered media to visit units in their areas of operation. The MOC will contact a unit and coordinate the time and location for the visit, as well as notifying a PA or nonPA escort for a mission. The escort will need to know the unit's location and whether the unit 's UPAR (if applicable) will travel with the group or meet them at the unit. The unit will need to know the number of media visiting, the duration of the visit and the status of the unit for the last 24 hours (i.e., has the unit been in contact with the enemy and has there been any significant actions).

6-29. The escort will plan the route to the unit. The number of media going will determine the number of escorts and vehicles needed for the trip.

6-30. **Preparation**. The JIB or escort will notify the media and tell them the time and location of departure and estimated time of return. This allows the media to plan for their supplies and equipment for the trip. The escort should meet with the media and update them on the unit's status, find out if the media has any specific requests before departing, cover safety points and OPSEC requirements, and ensure they have everything they need. This will also provide the escort with a direction for the types of questions or issues the media may address. The information concerning the unit's situation will allow the escort to develop a list of Q&As for the commander and individuals whom the media will interview.

6-31. The escort should meet with the drivers and review the route to the unit, cover contingency plans and determine an inspection time before movement. The escort should also contact the unit to reconfirm the visit and their location, as well as providing a list of the names and agencies of media visiting them. The escort should meet the media at the vehicles to insure they have what they need for the trip.

6-32. **Execution.** The escort should leave as scheduled and travel as the tactical or non-tactical situation dictates.

6-33. When arriving at the unit, the escort should have the driver park the vehicle as directed by the unit and have media personnel wait at the vehicles (if possible, with military supervision). The escort needs to request that the media not take photographs of the unit's position for operational reasons. Let the media know you are going to get the commander and will return as soon as possible. If the situation is hostile, get the media inside the perimeter and secure them away from equipment and the command post.

6-34. The escort should meet with the UPAR and let him/her know who is in the media group. Ensure that the unit will be able to provide the interviews the media requested.

6-35. The escort should brief the commander and let him know the ground rules, covering possible Q&As. Answer any questions or concerns he may have about the interview. Let him know that you will be present during the interview to assist him. When possible and if mission requirements permit, the commander should be available for the media. Review the unit's OPSEC requirements to ensure you understand what cannot be photographed.

6-36. The escort should determine where the commander wants to meet the media. If the area is not appropriate for conducting interviews (may be based on PA experience or media request), provide an alternative recommendation to the commander.

6-37. The escort should move the media to the interview point and introduce the commander to the group. After the interview, the escort should facilitate the rest of the coverage with the help of the UPAR. Do not allow the media to linger in the area of operations and become mission detractors. Wrap up the visit, return to your vehicles and depart. Upon returning to the JIB, the escort should be available to back brief the PAO.

6-38. The requirement to maintain proficiency in the full range of public affairs collective and individual capabilities and skills places a high priority on tactical training for both AC and RC PA soldiers. Tactical training participation enhances knowledge of battlefield requirements, increases unit cohesion, and forms the basis of experience needed for operational planning, mobilization, deployment and mission success.

6-39. PA will continue to face expanded missions in the joint arena. Quality PA training provides sufficient numbers of trained PA personnel to conduct joint and combined PA missions. Training must emphasize the joint perspective, enhance interoperability, and contribute to each combatant CINC's PA Mission. Exercise participation is critical to the training of PA personnel.

6-40. Trainers must fully incorporate a broad array of PA activities into all types of exercises and war gaming. These activities can be injected into computerized battle simulation as the training exercise driver. They should also be used in BCTP command post exercises, and JRTC, CTC, and NTC rotations. Seminars, area assessments and TEWTs are all forms of training that also provide relevant, realistic training.

Chapter 7

Community Relations

Public opinion about the Army is greatly influenced by the actions of each command. What the command does for its local community or fails to do affects the perceptions and attitudes of the American people, upon whom the Army depends for its support and existence. This applies not only to official acts but also to unofficial acts, which by their commission or omission affects public opinion. This principle also applies to individual members of the Army, their dependents and Army civilian employees in their personal contacts with the civilian community. Conducting community relations is a vital element to successful public affairs operations. Commanders and public affairs officers (PAO) must seize on key opportunities to gain and maintain links to internal and external publics.

EFFECTIVE COMMUNITY RELATIONS

7-1. PA personnel act in concert with veteran's groups, civic leaders and local populations to increase understanding and build support for Army activities. Army support to and participation in public events is based on the fact that the Army belongs to the American people. Common ownership requires that Army resources be used to support events and activities of common interest and benefit.

7-2. Effective community relations requires:

- Command supervision at all levels.
- Appreciation of public opinion and attitudes toward the Army
- Planning definite actions and positive policies.
- Implementing programs in a competent, professional and responsible manner.
- Constant evaluation of continuing programs to measure their effect upon the public and the command.
- Sharing the results of the program.

7-3. Commanders must maintain continual liaison with persons and organizations in the local community to help resolve common problems and develop cooperation and understanding between the installation and the local community. Community relations develop an effective two-way channel of communication between the Army and the community. PA does this by capitalizing on opportunities for better relations and resolving potential and actual areas of conflict.

7-4. Community relations projects or programs may be supported by use of exhibits, equipment and facilities. Exhibits consist of displays such as mission exhibits models, devices and other information and orientation materials at conventions, conferences, seminars demonstrations, exhibits, fairs or similar events. Also included are exhibits displayed on military installations during open house programs.

COMMUNITY RELATIONS ACTIVITIES

7-5. The goals of community relations is to develop an open, mutually satisfactory, cooperative relationship between the installation and the community. A successful community relations program improves the community's perception of the Army and its appreciation for the installation and the soldiers, family members and civilians who are part of the installation. It is based on openness and honesty. Community relations objectives are community assistance and social improvements for the community in which the military must work and live.

Historical Perspective

Fort Eustis, Va. started a pilot program in 1992 called Operation Self-Enhancement to give high-risk middle school students the opportunity to visit the post and focus on careers, teamwork and self-esteem. The program was so successful that it has become an annual event. Students receive light military training through an array of testable tasks and obstacles presented by members of a cadre team. This training helps students build their self-esteem and self-confidence and affords them the opportunity to interact with positive role models. This also gives the students a better idea of what the Army is about.

7-6. Community Relations activities include:

- **Speakers Bureaus.** Speakers are an effective means of developing understanding of the Army, stimulating patriotic spirit and informing the public about the activities of the installation, its units and its soldiers. Commanders should establish an installation speakers bureau and encourage soldiers of all ranks to participate in the installation program.
- **Community Liaison.** Maintaining liaison through informal community relations councils can establish and maintain open communications with community officials and organizations. Councils can be charged with a variety of responsibilities, such as developing and promoting new ways for members of the command to actively participate in local community activities, capitalizing on opportunities for better relations and resolving potential and actual areas of conflict. Community liaison can also involve recognition of private citizens, local community leaders and citizen groups and organizations for their support of the Army by public service awards. Commands can further community liaison through membership in civic, business and professional organizations when

the goals and objectives of those organizations are beneficial to the Army and their programs and projects are consistent with Army interest.

- **Ceremonial Units.** The band, color guard and other ceremonial units participating in public events are excellent ways to accomplish community relations objectives. These representatives of the Army serve as ambassadors to the civilian community and promote patriotism, interest in the Army, and awareness of the professionalism of our forces.
- **Exhibits.** Exhibits and displays of Army equipment, historical materials, models, devices and other information are other community relations activities that can enhance understanding of the Army and the installation. They provide an excellent opportunity for interaction between our soldiers and members of the local community and can communicate the professionalism, readiness and standards of our forces.
- **Open House.** Open houses may be scheduled to coincide with Armed Forces Day, the Army Birthday, service branch birthdays or anniversaries which mark the history of the installation, a unit or community events or in support of media day. An Open House gives the local community an idea of who we are and what we do. They also have the opportunity to visit us on the installation-- at our job site.
- **Physical Improvements**. Community service physical improvements focus on ensuring that the physical infrastructure is as safe as possible and provides the fullest possible range of support to the population. These activities encompass a wide range of programs that do not compete with the services provided by contractors and businesses in the local civilian community.

7-7. Some examples of physical improvements are:

- Construction projects that enhance the recreational, educational, environmental or cultural facilities of the community, such as building community picnic areas and hiking and biking trails.
- Demolition projects that enhance the safety and appearance of the community, such as the removal of unstable playground equipment.
- Projects that create or enhance a safe, clean environment, such as the removal of debris from a community wildlife area or painting a community recreation center.

TOWN HALL MEETINGS

7-8. Town Hall meetings provide installation commanders with an unfiltered means of communicating ideas to internal and external communities. This tool for conveying important information and ideas about the command cannot be underestimated in its effect and should not be planned haphazardly.

7-9. Commanders, PAOs and staff directorates must work together to produce an effective community relations product.

7-10. Prospective town hall meeting planners must understand, and properly apply, the correct type of town hall meeting. With a focus on the type of meeting and probable audience, the planner can begin the process of planning and conducting the event.

7-11. As part of the plan, the planner must determine the likely audience for the meeting, including attendees from internal and external audiences. He must also evaluate possible attendance by key publics. The planner should develop a standing operating procedure (SOP) to ensure each mechanism of the process is in place for the scheduled event.

7-12. Finally, post-event analysis is imperative to accurately assess its effectiveness.

7-13. The PAO must develop systems to quickly assess the feedback data and activate a follow-up plan that will maintain confidence from the community that town hall meetings are meaningful events.

TYPES OF TOWN HALL MEETINGS

7-14. Installation commanders can stage various types of town hall meetings. The commander must determine which meeting type, or hybrid, is appropriate for disseminating information and gaining useful feedback from internal and external publics.

7-15. The following meeting program structures have inherent strengths and weakness; knowing the potentials for message delivery will assist the commander in making his decision:

- **Commander - Expert Format**: This meeting (Figure 7-1), is characterized by the commander attending with key staff members facing a live audience. Typically the commander and his staff will give presentations and then field questions from the audience. Usually, attendance is open to the public.
- The primary advantages to this format center on the open nature of the meeting.
- This meeting provides the commander with an opportunity to provide detailed presentations with time being a minimal constraint. He also receives instant feedback from the types of questions from the audience and the passion with which questions are asked.
- Meetings in this format are likely to be seen as the most easily accessed by the internal and external publics.
- Among the disadvantages of the format is the ability to reach large audiences and control the conduct of the meeting.
- Unless the meeting is taped for later airing on the commander's cable access channel, the audience is often narrowly focused.
- Although the possibility exists that large audiences will attend, it is also possible small or narrowly focused audience will limit the general effect the commander seeks.
- An open meeting can also become the forum for unruly or disgruntled audience members to incite others or attempt to draw the commander into an open confrontation.
- This factor can be mitigated through the use of question time-limits and use of a moderator (other than the commander), but cannot be

totally eradicated. (As with any other public event, security must be a consideration in the planning process).

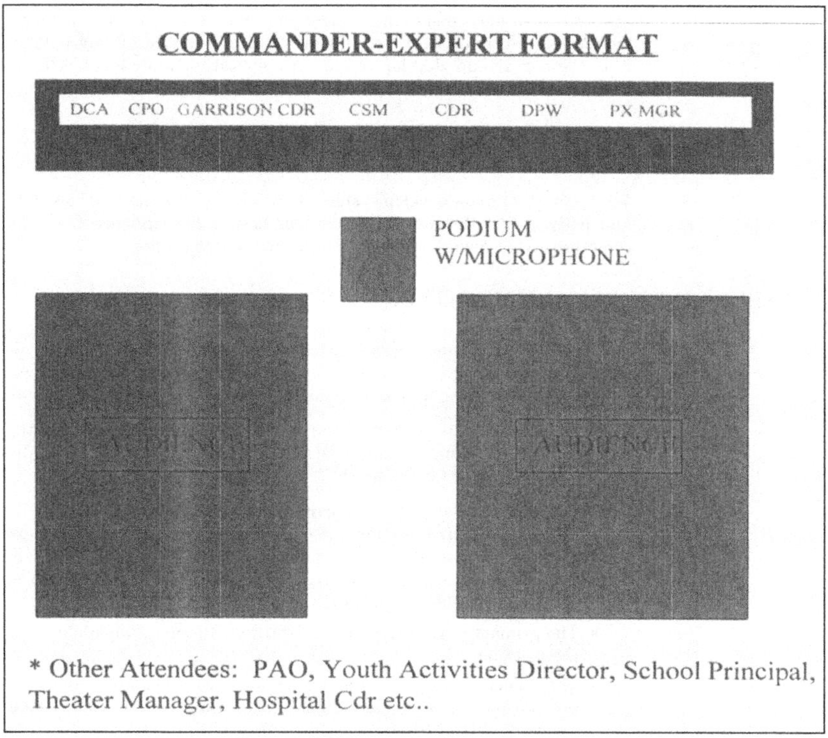

Figure 7-1. Commander Expert Format

7-16. Commander Access TV Channel Format: This meeting format (Figure 7-2) uses the commander's cable access channel to air the event. Normally, the commander, CSM and Garrison Commander (and other staff members as necessary) give a presentation. No live audience is in attendance. The commander provides a set of phone numbers, allowing questions to be called in. The staff operates the phone bank, accepts the questions, directs the questions to the appropriate staff agency, and delivers the answers by 3x5 card to the on-air panel. The panel members read the questions/answers to the viewing audience.

7-17. This format offers advantages focused on control and distribution of the product. The venue and the setting are completely controlled by the commander and his staff. The staff screens questions and, thus, no surprises will occur. Indeed, if questions/answers are given to the commander, he may choose to either not answer or return the card for more information.

7-18. Other advantages include an ability to re-run the meeting as often as desired and provide copies to local cable providers, many of which are interested in using the product for airing.

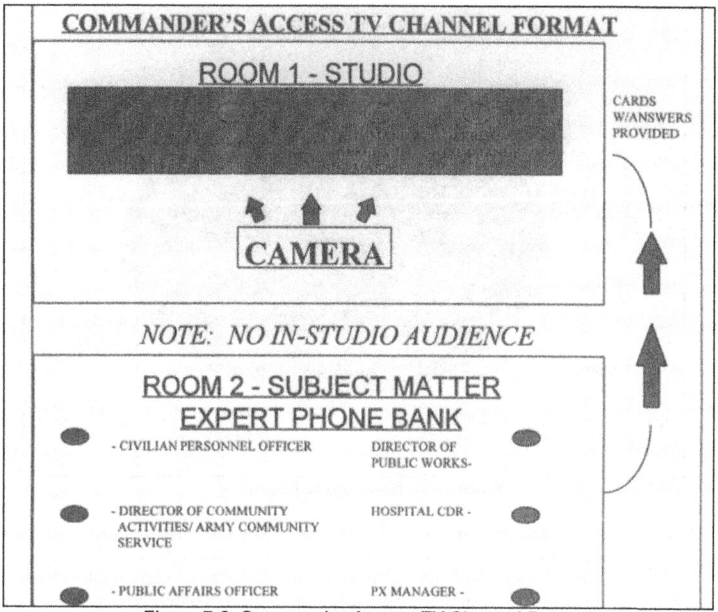

Figure 7-2. Commander Access TV Channel Format

7-19. Disadvantages to this format center on a difficulty in gauging feedback, competing for viewing audiences on television, and providing personal contact with the commander. With no attending audience, collecting feedback data is problematic. Follow-up questionnaires, using statistical probability methods, provide the only means of gaining reliable feedback. The likelihood of viewers to "channel-surf" is high due to the specific nature of questions from callers. Holding the interest of viewers, all of whom have multiple other viewing options available, is difficult. Finally, by appearing on television, in uniform, with a phalanx of staff members, the commander risks appearing to be speaking from "the mount." Audiences may view the commander as speaking down to them and being out-of-touch to their concerns.

7-20. Key members from the internal and external publics form a roundtable discussion (Figure 7-3). The topics are set by published agenda, with some time left for open discussion. The meeting results are published for general distribution. Media are usually invited as participants.

7-21. Roundtable discussions with key publics offer advantages to the commander by providing information to the individuals who represent overlapping and wide constituencies. By setting an agenda, the commander can deliver focused messages with a high likelihood of the messages later reaching targeted audiences. It allows the commander and potentially key staff officers to deliver presentations and provide follow-up information to those in attendance.

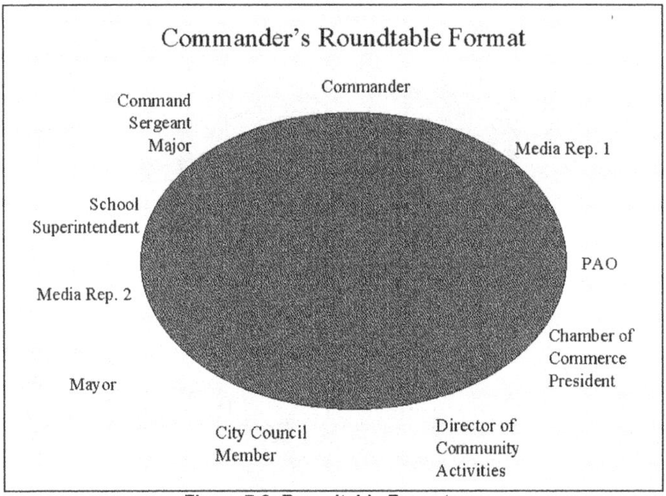

Figure 7-3. Roundtable Format

7-22. The principal disadvantage is the distance between the commander and the publics. The publics will receive information about the meeting second-hand and in a potentially filtered state. Gaining feedback may also prove challenging (but not necessarily impossible). Other disadvantages may be in the limits of the audience makeup; key publics, which are not invited, may resent the slight.

7-23. Commander-in-the-round format. A room is set-up that will allow the commander to be in the middle of all the attendees. Normally, he will stand and walk around in the circle formed by attendees. The meeting is usually open to the public.

7-24. The primary advantage to this format is the close contact the commander shares with his audience. Audience members may feel that barriers are lowered because the commander is close in proximity and no staff members are buffering their access. The commander also can realize feedback very quickly and can gain some appreciation for the resonance of his ideas with the assembled audience.

7-25. Such closeness with the audience can also be a significant disadvantage. Limiting surprises and controlling potentially unruly audiences is extremely difficult in this format. Further, because of the commander's reliance on his personal notes and memory, his ability to provide detailed information and multi-media presentations may be restricted.

7-26. Mitigation against such limits is dependent on the site design. Other disadvantages include problems associated with other "open" formats, including a possibly non-representative makeup of the audience.

7-27. Before deciding on which format is appropriate for the information, the commander must have a plan for what must be achieved.

7-28. Characteristics of an effective plan:

7-29. Answers the "why?" questions. The planner must understand the purpose of the meeting to correctly advise the commander on the format and substance to begin coordination of the plan. As described above, the various formats each have strengths and weaknesses that will assist the planner to shape the meeting.

7-30. Routinely Scheduled. Normally, meetings can be scheduled quarterly or monthly. Use the installation planning cycle to ensure proper coordination and notification of public meetings.

7-31. Site Plan. Checking and securing a site for the town hall meeting is dependent upon the type of meeting selected. Plan for the site early in the process and establish the layout of the site in detail.

7-32. Calendar Check. Before scheduling events, check local and regional calendars for possible conflicts. For example, scheduling a meeting on the same night as the local high school homecoming football game may prove disastrous.

7-33. Presentation Submission Deadline. Coordinate with appropriate staff agencies responsible for preparing presentations. Ensure presentations are properly staffed and approved. As PAO, establish a firm submission cut-off date.

7-34. Focused Presentation. Inform the staff of the commander's intent for the meeting. If the commander wants a particular theme addressed, ensure staff agencies adhere to the parameters of the intent.

7-35. Media Invitation List. Invite local and regional media including print and electronic outlets. Develop relationships with individual reporters and provide background material as necessary. (Local newspapers, radio stations and television stations can often assist in publicizing meetings as well).

7-36. Publicity Plan. Ensure all available avenues are used to publicize meetings. Included in this process are the post newspaper and radio station (where available), normal distribution, staff meetings, E-mail delivery and chain of command communications. Take special care to invite key publics by individual invitation and phone call follow-up.

EVALUATING FEEDBACK

7-37. Assessing the effectiveness of town hall meetings is essential to developing community-related policies and courses of action for the command. PAOs can use standard statistical measurements using survey techniques to gauge the level and intensity of views of the various publics. Other analyses can be derived from follow-up media content analysis, letters to the commander and post newspaper, and reactions at the meetings. None of the methods described here will render a perfect picture. PAOs must exercise good judgment and personal insight when advising the commander of analytical results.

7-38. **Response Follow-up**. Investigating and responding to issues raised at town hall meetings are critical to public perceptions of the level of care the commander applies to community operations. Just as a maneuver commander sees battle damage assessment (BDA) as crucial to determining the efficacy of fires, commanders and PAOs must determine the true productiveness of community relations programs. Commanders will often feel compelled to promise action, such as investigation or immediate problem resolution, during a meeting. Staff agencies are normally the conduit for actions (only occasionally will the commander personally provide the requisite service).

7-39. PAOs should provide the staff oversight of the response mechanism. PAOs can devise a recurring memorandum that provides the commander (usually through the chief of staff), details about the status of actions. All staff agencies should receive updates routinely.

7-40. Along with the response mechanism, PAOs can use command information and media relations channels to inform the publics of problem resolutions. For example, if an issue raised at a town hall meeting indicated that the local recreation center was routinely opening two hours late each day, describe the measures taken to alleviate the problem. The PAO can use the post newspaper or radio station to provide lists of problems and resolutions.

CONCLUSION

7-41. Community relations and activities are vital to instilling and maintaining the confidence of internal and external publics in our great Army. Commanders and PAOs cannot leave the prospect of successful relations to chance. Too much is at stake. Careful selection of the type of town hall meetings to be used must be taken. The event must be planned with the attention to detail required for all military operations. Each part of the plan has unique importance and cannot be overlooked. To ensure the effectiveness of the operation, PAOs are compelled to build a clear mechanism for evaluating outcomes. Finally, the command must provide conspicuous follow-up responses to issues raised to complete the process. Town hall meetings provide an excellent opportunity for commanders and the various publics to interact and improve community institutions.

COMMUNITY ASSISTANCE

7-42. Community assistance applies the skills, capabilities and resources of the needs and interests of the local community. Providing support for and participating in events and activities which are beneficial to both the Army and the community, builds on a long tradition of "America's Army" helping American communities. Identifying opportunities, which advance the interests of both the Army and the community, is an important objective for every commander.

7-43. Community assistance projects and operations must impact positively on the unit or individual soldier, enhance unit or individual readiness and contribute toward the common good of the community. Army commanders must ensure that their initiatives are not competitive with local resources or services and do not benefit any particular interest group and will not result in monetary or service remuneration in any form for unit members or the unit as a whole.

7-44. Increasing public awareness and understanding of the Army, inspiring patriotism and enhancing the Army's reputation as a good neighbor is a goal of community assistance. Community assistance activities can help build unit morale and esprit de corps. These activities also provide an excellent opportunity for soldiers to serve as role models, which not only enhances recruiting efforts, but also serves to motivate soldiers by promoting their self-esteem and furthering their sense of service to the nation.

READINESS ENHANCEMENT

7-45. Certain community assistance activities enable a commander to train soldiers, enhance individual and unit readiness, maximize use of assets and foster a positive training environment where soldiers can become involved in realistic, "hands-on" training opportunities. Projects should be selected which exercise individual soldier skills, encourage teamwork, challenge leader planning and coordination skills, and result in measurable, positive accomplishment. Finally they should enable a unit to exercise use of its equipment, resulting in training opportunities that can generate greater operator efficiency for future missions.

7-46. Examples are:

- Medical Readiness Program. The Medical Readiness Program is an activity in which Army medical unit personnel, together with state medical emergency officials, plan and provide support in the form of diagnosis, treatment, and preventive medical, dental, and veterinary care to citizens in remote areas of the U.S. or its territories. The program is designed to enhance the unit's medical readiness, provide unit training opportunities and serve the public in locations where medical care is not otherwise available. The program may not compete with local private medical care that may be available.
- Air Ambulance Participation. The Military Assistance to Safety and Traffic Program. (MAST) is a proven example of Army support to civil authorities. This program permits the utilization of Army aviation assets to conduct emergency air evacuation and recovery efforts.
- These projects contribute to the health and welfare of the community, making the Army an integral partner in community progress and development. They can enhance the ability of the local community to support itself and its people, to provide the best possible services to the citizens and to promote a positive, healthy safe environment.
- Community service activities are those which focus on improving the community, its infrastructure and its ability to serve the local population.

7-47. **Social Improvements.** Community service social improvements, which focus on making the social environment as healthy as possible, provide the widest range of support to the population. They encompass many projects including:

7-48. Support to youth programs, such as scouting and programs that provide assistance to special need audiences such as the Special Olympics.

7-49. Examples are:

- The Civilian Youth Opportunities Program (Challenge). This is a youth program directed at attaining a high school diploma, providing job training and placement, improving personal and social skills and providing health and hygiene education and physical training. Soldiers work with civilian leaders to provide a comprehensive support package, ranging from choosing appropriate clothing to attending residential training facilities.
- Youth Physical Fitness Clinic Program. The National Guard encourages fitness and combines academic and athletic achievement by helping schools conduct competitions in selected athletic events. This program also establishes a separate scholar-athlete category for those students with a 3.5 or higher grade point average.

7-50. Involvement in ventures and projects that enhance the educational or cultural climate of the community, such as adult literacy, reading or community theater programs.

7-51. Examples are:

- Civilian Community Corps. This program provides managerial, organizational and technical skills for disadvantaged Americans seeking the skills for success. Through this program, the Army helps participants become productive citizens. In exchange, participants perform a wide range of community service activities that improve the foundation of American society. This program encourages intra-governmental cooperation on the federal level. It also encourages partnerships with industry, education, state, federal and local governments.
- Science and Technology Academies Reinforcing Basic Aviation and Space Exploration Program. (STARBASE) This program is an innovative partnership of professional educators, military personnel and corporate sponsors. It promotes science, mathematics and technology basics for primary through secondary schools. Using National Guard resources to spark student interest, the program develops strong self-esteem, provides excellent role models, promotes positive attitudes and develops goal-setting skills.
- The guiding principle behind community relations efforts is that the installation and the community have a common interest in providing the best possible support for each other. A cooperative relationship exists, because soldiers stationed at the installation receive life support from the community while many of the civilians who make up the community receive life support from the installation. The interdependence of the military installation and the civilian community can involve economics, education, health care, basic services, quality-of life issues and many others.
- The community relations goal of local commanders is to develop an open, mutually satisfactory, cooperative relationship between the installation and the community. These efforts improve the community's perception of the installation and the soldiers, family members and civilians who are part of the installation.
- Participation in community relations activities is an effective method for projecting a positive Army image, making the best use

of assets, providing alternative training opportunities and enhancing the relationship between the Army and the American public. Activities vary widely, ranging from individual soldier involvement to full Army participation. They are characterized by detailed coordination between the military command and community authorities. They fulfill community needs that would not otherwise be met, enhance soldier and unit morale, skills and readiness and improve the mutual support between the military and civilian communities.

Chapter 8

PA Organizations

THE BRIGADE

8-1. The Brigade PAO is the lowest level to which the Army has assigned organic Public Affairs Assets.

8-2. Working as both a special staff officer and as a member of the Brigade's planning team, the BDE PAO acts as the spokesperson for the unit, advisor to the Commander, and provides Public Affairs guidance and planning to commanders at all levels.

8-3. To support and conduct Public Affairs Operations within the Brigade, the BDE PAO has the following functions, organization and equipment:

8-4. Specific functions of the Brigade PAO, as outlined within the core competencies are:

- **Public Affairs Planning**
  - Advise the commander and staff on PA implications of plans and actions.
  - Write Brigade PA annex and matrices using operational tools, terms, graphics, and concepts
  - Execute the plan
  - Plan for future operations.
  - Command and control attached Public Affairs assets within the brigade AO.
  - Tactically communicate with PA units and supported combat units in the Brigade's Battlespace.
- **PA Training**
  - Train and supervise stringers to assist their commanders in conducting their internal information programs.
  - Train soldiers, family members, DACs and everyone habitually associated with the unit to comfortably and successfully communicate with the media.
  - Utilize Train the Trainer.
  - Evaluate public affairs training programs of subordinate commands.
  - Train subordinate commands to facilitate the media.

  - Train leaders and soldiers to protect information products and information systems from compromise and intrusion by practicing security at the source.
- **Media Facilitation**
  - Assist media to gain access to units, soldiers and commanders.
  - Evaluate subordinate command's media facilitation plan.
  - Assist subordinate commanders to coordinate transportation.
  - Conduct situation briefs as needed.
  - Monitor media operations within the Area of Operations. (Maintain accountability of the media to preclude fratricide. Accomplished through reports from escorts or through electronic tagging and monitoring.)
  - Respond to and mediate potential media problems; react quickly to coordinate and facilitate information issues in Brigade's Battlespace.
- Information Strategies
  - Monitor and analyze the local Military Information Environment.
  - Provide public affairs support to the G5 / S5 for the development and implementation of civil affairs programs.
  - Support higher echelon Public Affairs requirements for information.
  - Gather Open Source Information to help build relevant information.
  - Assist Commanders to conduct internal and external information to include Hometown News Release Program.
  - Monitor local news media products (Visual, print and audio) and analyze for PA implications.
  - Monitor and Analyze the local Military Information Environment.
  - Protect digital images, information products and PA and non-PA information systems from compromise and intrusion.

8-5. Organizational Structure:

- Captain , 46A.
- Sergeant, 46Q

8-6. Equipment:

- Computer with compatible software and communications hardware
- Access to the Tactical Internet and tactical battlefield radio communications
- Access to FM, Satellite, and video and cellular communications
- Access to Army Battle Command System, (MCS\P or CSS\CS)

8-7. Transportation:

- HMMWV

8-8. Addititional Support Requirements. PAO requires linguistic support from the Civil Affairs soldiers or contract civilians attached to the Brigade.

SAMPLE PA EXECUTION MATRIX BRIGADE PUBLIC AFFAIRS OFFICER

	PHASE I	PHASE II CROSS LD	PHASE III DEFEAT	PHASE IV SEIZE OBJ. LION	PHASE V RECONSIT.
BRIGADE	PAD CIVILIAN MEDIA				PAD AND CIVILIAN MEDIA
1-66 AR	INTERNAL	EMBEDDED MEDIA-- PRIORITY OF EFFORT			EXTRACT MEDIA
1-4 AV	AVIATION WEEKLY	SAME	SAME	SAME	AVIATION WEEKLY DEPARTS
1-5 IN (L)	OWN THE NIGHT MAG.				
SCOUTS		MEDIA REPS LINK PRIOR TO DEPLOY			
BSA	FOOD PREP QTRLY	SAME	SAME	SAME	SAME
MEDICAL	SAFETY STORIES				MEDIA - INJURED SOLDIERS

THE DIVISION

8-9. The Division PAO is the next level to which the Army has assigned organic Public Affairs assets. The division is largely self-sustaining and capable of independent operations. The division is a unit of maneuver, organized with varying numbers and types of combat, combat support (CS) and combat service support (CSS) units.

8-10. The division may be armored, mechanized, medium, light infantry, airborne or air assault; each can conduct operations over a wide range of environments.

8-11. Working as both a special staff officer and as a member of the division planning team, the DIV PAO acts as the spokesperson for the division, advisor to the Commander, and provides Public Affairs guidance and planning.

8-12. To support and conduct Public Affairs Operations within the division, the DIV PAO has the following functions, organization and equipment:

8-13. Functions: Assumption #1: DMAIN is not in country

- Gather, analyze and disseminate open source information, focusing on global media, to increase the situational awareness throuhout the command.
- Gather and disseminate multimedia products throughout the command, and to external and home station audiences.
- Conduct Public Affairs research and write the PA Estimate of the situation.
- Write the PA plan for OPORDERS, OPLANS, and TACSOPS.
- Monitor and analyze American and foreign public sentiment of current operations from available media sources for PA implications and advice.
- Monitor and analyze battlespace communications (visual, audio, FM, satellite) for PA implications.
- Monitor news media products (Visual, print and audio) and analyze for PA implications.
- Facilitate Media Operations.
- Subfunctions:
    - Assist media to gain access to units, soldiers and commanders.
    - Coordinate air and ground transportation on a non-interference basis.
    - Conduct daily situation briefs as needed.
    - Respond to media queries.
    - Validate media's credibility, expertise, knowledge, purpose and legitimacy.
    - Train and supervise unit-level PA representatives (Command Information NCOs) to assist their commanders command information programs.
    - Train soldiers, family members, DACs and everyone habitually associated with the unit to comfortably and successfully communicate with the media.
    - Survey soldiers, family members, DACs and other members of the internal audiences to measure effectiveness of the command information program.

8-14. Assumption #2: DMAIN is in country

- Advise the commander and staff on PA implications of plans and actions
- Gather, analyze, and disseminate Open Source Information, focusing on global media to increase the commander's situational awareness.
- Gather and disseminate multimedia products throughout the command, and to external and home station audiences.

- Conduct Public Affairs research and write the PA Estimate of the situation.
- Write the PA plan for OPORDERS, OPLANS and TACSOPS
- Monitor and analyze American and foreign public sentiment of current operations from available media sources for PA implications and advise
- Monitor and analyze battlespace communications (visual, audio, FM, satellite) for PA implications
- Monitor news media products (Visual, print and audio) and analyze for PA implications
- Coordinate and integrate all information-related functions (PSYOPS, CA, VI, Joint, Combined and Interagency PA) into the PA plan
- Conduct Information Operations
- Act as a conduit for CI products from the field, sanctuary and commercial sources for input into the commander's information program. Provide command information to soldiers, family members and Department of the Army Civilians.
- Facilitate Media Operations
- Subfunctions:
  - Coordinate air and ground transportation on a non-interference basis
  - Assist filing stories, video and photographs on a non-interference, reimbursable basis
  - Conduct daily situation briefs as needed
  - Respond to media queries
  - Validate media's credibility, expertise, knowledge, purpose and legitimacy
  - Conduct Primary and Secondary accreditation
  - Primary--Full accreditation of non-accredited media
  - Secondary--Process media previously accredited at Corps and higher
  - Survey soldiers, DACs and other members of the internal audiences to measure effectiveness of the command information program

8-15. Organizational Structure:

- At a minimum an embedded PA division section has:
  - Major 46A and two Captains 46A
  - Master Sergeant 46Z
  - Specialist 46Q
  - Specialist, 46R
  - PFC 46Q

8-16. Equipment:

- Computer with compatible software and communications hardware and tactical fax machine.

- Access to the Tactical Internet and tactical battlefield radio communications (SINGARS, MSRT etc.)
- Access to FM, Satellite, video and cellular communications
- Access to Army Battle Command System, (MCS\P)

8-17. Transportation:

- Light Division--HMMWV
- Armor and Mech Division-- HMMWV and Trailer

8-18. TAC 1 AND 2 Missions and Functions

TACTICAL COMMAND POST PA SECTION

8-19. The TAC CP, Public Affairs Section is task-organized based on METT-TC. It is the center of gravity for immediate internal and external communication, resolution of Public Affairs issues and violations of ground rules by media representatives. It acts as a conduit to the Information Operations Cell in the DMAIN adding immediacy to the Open Source Information process.

8-20. Public Affairs personnel deploy with the lead elements during any contingency and remain with the TAC CP to assist the commander, provide internal command information to deployed soldiers and limited external information to home station. They conduct media facilitation to expedite the flow of information to the America public while freeing the commander and his soldiers to conduct their mission.

8-21. The TAC CP, Public Affairs Section maintains connectivity with commanders and global information sources. It contributes to and monitors the common relevant picture, and synchronizes collection and dissemination efforts of soldiers far forward to internal and external audiences.

8-22. Particularly during split-based operations, the PAO task organizes his section to best serve the commander, his soldiers and the American public's need for information.

8-23. Assumption #1: DMAIN is not in country

- Act as the division spokesman.
- Advise the commander and staff on PA implications of plans and actions.
- Execute the plan.
- Assist DMAIN to gather Open Source Information to build the common relevant picture.
- Conduct Information Operations.
- Gather and disseminate multimedia products throughout the command and to external and home station audiences. Act as a conduit for CI products from the field, sanctuary and commercial sources for input into the commander's information program.
- Monitor and analyze battlespace communications (visual, audio, FM and satellite) for PA implications.
- Monitor local news media products (Visual, print and audio) and analyze for PA implications.

- Coordinate, integrate, and synchronize all information-related functions (PSYOPS, CA, VI, Joint, Combined, and Interagency PA).
- Synchronize Public Affairs assets (internal and external) within AO.
- Facilitate Media Operations:
  - Assist media to gain access to units, soldiers and commanders.
  - Coordinate transportation on a non-interference basis.
  - Assist filing stories, video and photographs on a non-interference, reimbursable basis.
  - Conduct daily situation briefs as needed.
  - Respond to media queries.
  - Monitor media operations within the AO.
  - Respond to and mediate potential media problems; react quickly to coordinate and facilitate information issues throughout the AO.
  - Validate media's credibility, expertise, knowledge, purpose and legitimacy.
  - Provide seamless connectivity for media accountability throughout the AO.
  - Conduct Primary and Secondary accreditation.
  - Primary--Full accreditation of non-accredited media.
  - Secondary--Process media previously accredited at Corps and higher.

8-24. Assumption #2: DMAIN is in country

- Act as the division spokesman.
- Advise the commander and staff on PA implications of plans and actions.
- Execute the plan.
- Assist DMAIN to gather Open Source Information to build the common relevant picture.
- Monitor battlespace communications (visual, audio, FM and satellite) for PA implications.
- Facilitate Media Operations:
  - Assist media to gain access to units, soldiers and commanders.
  - Conduct daily situation briefs as needed.
  - Monitor media operations within the AO.
  - Respond to and mediate potential media problems; react quickly to coordinate and facilitate information issues throughout the AO.
  - Provide seamless connectivity for media accountability throughout the AO.
  - Conduct Secondary accreditation.
  - Secondary--Process media previously accredited at Corps and higher.

8-25. The division public affairs section provides public affairs support to the division commander and to divisional units deployed in support of combined or joint operations. The division PAO has operational and tactical control over all PA TOE organizations assigned or attached to the division and coordinates closely with embedded PA sections within brigades or divisions to carry out PA operations.

8-26. The division PA staff, when deployed, is augmented by one PAD and one MPAD per three combat brigades. The division public affairs section, if augmented only by a Public Affairs Detachment, operates the division media operations center.

8-27. Traditionally, divisions have operated as part of a corps. In corps operations, divisions normally comprise 9 to 12 maneuver battalions, organic artillery battalions and supporting CS and CSS units. Divisions perform a wide range of tactical missions and for limited periods are self-sustaining. Corps augment divisions as the mission requires.

Chapter 9

Information Operations

INTEGRATING PUBLIC AFFAIRS

9-1. Information Operations involve a variety of disciplines and activities that range from electronic warfare and physical destruction through cyberwar and information campaigns. Public affairs is a related activity of IO, and contributes to overall operational success, both real and perceived.

9-2. Successful integrated IO requires coordination of themes, messages and activities in order to leverage the massing of information effects. When synchronized with other military operations, IO is a combat multiplier.

9-3. Information campaign objectives cannot be neatly divided by discipline, such as PA, CA and PSYOP. The responsible organization cannot be easily determined solely by looking at the medium, the message or the audience. For example, information about weapons turn-in policy and collection sites may be disseminated through a variety of means. This could include direct contact by Civil Affairs personnel with local populations and their leaders; PSYOP print and broadcast products; news releases, press conferences and other media facilitation coordinated by PA.

9-4. In accordance with joint doctrine (Joint Pub 3-61, Doctrine for Public Affairs in Joint Operations), public affairs are an operational function designed to contribute to the overall success of joint operations. For Public Affairs, the audience may be internal or external, but the objective is constant: Soldiers, participants and the public must understand objectives, motives and the nature, scope and duration of friendly actions. The relevant audiences important to the commander are not limited to soldiers and the American public, but are also international as well as local to the operation.

9-5. Synchronized information operations contribute to military campaigns in a variety of ways. These contributions may:

- Gain or sustain support for the U.S. or coalition position
- Reduce the need for combat forces
- Influence events with non-lethal means
- Counter propaganda and disinformatio
- Discourage adversary offensive operations
- Deter hostile actions

- Undermine illegitimate regimes
- Support the maintenance of coalitions

9-6. Information Operations during peacekeeping operations:

- Promote peaceful cooperation
- Lower friendly force requirements
- Counter propaganda and disinformation
- Reduce friction leading to hostilities
- Gain and maintain the initiative
- Shape opponent plans and operations

9-7. These goals may not be achieved solely by tactical level information operations, but rather, may be theater and national-level issues that are reinforced by tactical-level message dissemination. This requires horizontal and vertical integration of themes and messages to achieve a massing of information effects.

IO STAFF ORGANIZATION

9-8. Composition of the Information Operations battle staff/coordination council or other such element is flexible and tailored to the operation and desires of the commander. (See figure 9-1). Notional IO staff structures are included in FM 3-13 (100-6), Information Operations.

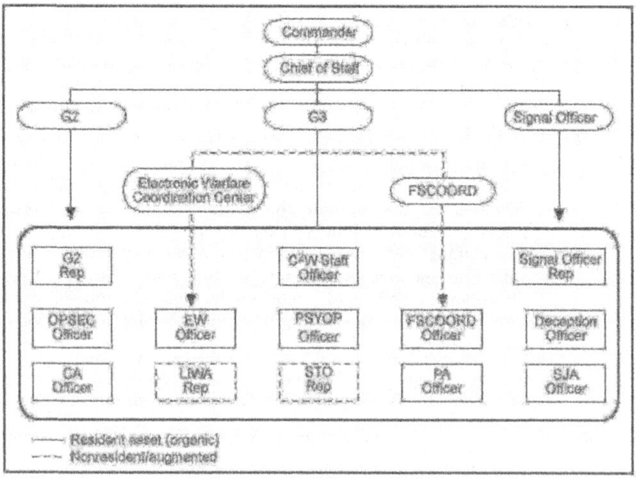

Fig 9-1 Notional IO Battlestaff

9-9. The IO cell is often headed by the G-3 or his designated representative, and includes representatives of a variety of organizations. The staff may include, but not necessarily be limited to the G2, G3, G5, G6, PAO, PMO, JAG, PSYOP, Electronic Warfare Officer, Political Advisor, Joint Military Commission representative, Fire Support Coordinator and Targeting Board representative.

9-10. The successful accomplishment of a specific mission may require the close coordination and synchronization of the range of information activities as well as maneuver elements. While the IO cell is lead by the G3 or his designate, IO coordination is the responsibility of the Land Information Warfare Activity field support team. While providing a public affairs representative to the IO Cell, PAOs must maintain a clear and direct link with the commander.

LIWA

9-11. The Land Information Warfare Activity, an INSCOM element, provides commanders with field support teams (FST) that serve as IO advisors in addition to effecting the synchronization and coordination of the range of activities that support IO. LIWA field support teams do not serve as functional area specialists, but rather, coordinate the activities of those elements. For example, the LIWA FST members may be from the military intelligence and PSYOP branches, but do not serve as the commander's intelligence analyst or PSYOP planner, or for that matter, Public Affairs advisor. They do, however, coordinate the actions and products of these and other activities in support of the IO plan.

PA SUPPORT TO INFORMATION OPERATIONS

9-12. PA participation in IO involves no completely new tasks but does require a broadened scope of operations. PA support to IO requires analysis of the Global Information Environment (GIE) and the operational environment, as well as synchronization of efforts with other organizations and agencies to ensure themes and messages are consistent and deconflicted.

9-13. PA in IO requires PA staffs to be fundamentally proactive rather than reactive. Often, actions may be taken and products developed to assist command achievement of a desired end state. This is more than merely reacting to events with a press release or conference

9-14. PA actions and events that support IO include print and electronic products, news releases, press conferences and media facilitation. PA advises the commander on how the operation is being perceived and portrayed and also provides guidance to unit commanders and soldiers. This includes regular talking points and themes for commanders and preparing soldiers to interact with the press. It's a means of emphasizing selected issues and positions--speaking with one voice.

PA ESTIMATE AND IO PLANNING

9-15. The starting point for PA contributions to Information Operations is the Public Affairs Estimate. (See Appendix C). The PA estimate consolidates information on the audiences, media presence, public opinion, personnel available and PA guidance.

9-16. This is not a static document created at the beginning of an operation, but must be continually updated to reflect changes in the operational situation and environment. Issues to consider include:

9-17. **Audience analysis.** Who are the audiences, both internal and external? What are their information needs? How do they get their information: television, radio, newspapers or word of mouth? Is the media state-run or independent? Does the audience population have telephones, cell phones, fax machines or Internet connections? These devices are frequently found even in developing countries and must be considered during the analysis of information channels.

9-18. **Media presence.** What media representatives and organizations are in the area of operation? Are they radio, television or print? Are they state-run or independent? What is their political slant? Are they pro- or anti-coalition? Are they receptive to coalition information products such as news releases or other print or electronic products? Is the local media interested in live interviews with U.S. commanders and soldiers?

9-19. **Public Opinion.** What are the opinions/beliefs of the local populations; of the international community; of the U.S. national population?

9-20. **Personnel available.** What is the available Public Affairs force structure (PADs, MPADs, BODs, CPIC/JIB staff, unit organic PA staff and individual additions), translators, Combat Camera and administrative staff.

9-21. **PA guidance.** What guidance has been received from higher levels? Official positions on theater issues are naturally not developed at the tactical level. What is the theater strategic/national command authority position? This is often coordinated and deconflicted at all levels via conference calls and other communication means.

IO CAMPAIGN CYCLE

9-22. There are four stages to an IO campaign cycle: capability development, assessment, planning and execution. The execution stage is accompanied by evaluation-- during and after the mission -- in order to adjust operations as needed and after the operation to gather lessons learned.

9-23. Capability development:

- Identify local resources and available external support. Theme and message delivery can take many forms, including radio/television, handbills, leaflets, loudspeakers, soldiers, displays, the Internet, internal information products, USIA, Voice of America, print and electronic news releases, press conferences, direct contact with parties, leaders, officials and citizens. Direct contact may include military liaisons, Civil Affairs personnel, diplomatic contact, or any form of personal interaction.
- Establish processes and procedures
- Collect, organize and store relevant information

9-24. Assessment

- Perform mission analysis
- Obtain commander's guidance
- Define IO goals and objectives
- Conduct risk assessment
- The assessment phase includes a mission analysis, clarification of the commander's guidance, initial identification of IO goals and objectives and a risk assessment. Goals and objectives include a

determination of the desired end state and what must be done to achieve it. This may mean inducing others to take or not take certain actions, or have the information to make certain decisions that will support the goals of the operation. Public Affairs is not in the business of shaping beliefs and attitudes of populations, but can provide factual information that enables people to make informed decisions.

- Risk Assessment
  - Consequences for command if information operation fails?
  - Potential unintended effects?
  - Operational success too reliant on IO?
  - Can IO campaign be used against U.S. or coalition?
  - Force protection issues?
  - Compromise or loss of impartiality?

9-25. Planning

- Develop and coordinate themes
- Determine the best implementation means
- Delineate tasks and responsibilities
- Identify feedback and measures of effectiveness channels
- Prepare implementation order
- During the planning phase, specific themes are developed and coordinated with all members of the IO cell. Message/theme delivery methods are determined and specific tasks and responsibilities are assigned. The IO cell may use a synchronization matrix to effectively manage IO events. For example, this matrix will indicate specific actions, events or products each member organization of the IO cell will execute or produce to support the plan. For example, a specific event may require PSYOP leaflets and broadcasts, PA press releases, news conferences and interviews with soldiers and Civil Affairs meetings with local officials and community leaders. These activities are coordinated on the matrix, ensuring deconfliction of resources, messages and products.
- Measures of effectiveness and feedback indicators vary widely and should be identified in the planning process. They may include questions raised by the media, editorials and commentaries, statements by public officials, postings to internet newsgroups and forums, demonstrations and protests, statements during meetings, responses given to public opinion surveys, behaviors during specific events, as well as other SIGINT and HUMINT collection and analysis.
- The product of the planning phase is a synchronization matrix and execution schedule. The matrix is then coordinated with the overall synchronization matrix, ensuring that IO is coordinated across the BOSs.

9-26. Execution

- Conduct the mission.

- IO monitoring must be conducted throughout the execution of the event and during follow-up review, feedback and evaluation.
- If necessary or possible, alter mission if evaluation determines it is not successful or unexpected responses occur.

9-27. **Evaluate**

- Assess the effectiveness of the operation
- Determine preventive methods, document lessons learned and apply to next operation

Appendix A

DoD PRINCIPLES OF INFORMATION

The DoD Principles of Information are contained in DoD Directive 5122.5, Change 1. They chart the course for all DoD Public Affairs activities, and apply to the full continuum of day-to-day activities and operations. It is the commander's responsibility to ensure that all planning for military activities and operations efficiently and effectively achieve the goals set by these principles.

DOD PRINCIPLES OF INFORMATION

A. Timely and accurate information will be made available so that the public, Congress, and the news media may assess and understand the facts about national security, defense strategy, and on-going joint and unilateral operations.

B. Requests for information from organizations and private citizens will be answered in a timely manner. In carrying out this policy, the following principles of information apply:

(1) Information will be made fully available, consistent with statutory requirements, unless its release is precluded by current and valid security classification. The provisions of the Freedom of Information Act and the Privacy Act will be complied with in both letter and spirit.

(2) A free flow of general and military information will be made available, without censorship or propaganda, to the men and women of the Armed Forces and their family members.

(3) Information will not be classified or otherwise withheld to protect the government from criticism or embarrassment.

(4) Information will be withheld only when disclosure would adversely affect national and operations security or threaten the safety or privacy of the men and women of the Armed Forces.

(5) The Department's obligation to provide the public with information on its major programs and operations may require detailed public affairs planning and coordination within the Department and with other government agencies. The sole purpose of such activity is to expedite the flow of information to the public; propaganda or publicity designed to sway or direct public opinion will not be included in Department of Defense public affairs programs.

Appendix B

THE GUIDELINES FOR COVERAGE OF DOD COMBAT OPERATIONS

DOD MEDIA GUIDELINES

The DoD Media Guidelines, issued as Change 3 to DoD Directive 5122.5, provide the following guidelines for coverage of DoD combat operations:

A. Open and independent reporting will be the principal means of coverage of U.S. military operations.
B. Pools are not to serve as the standard means of covering U.S. military operations. But pools may sometimes provide the only feasible means of early access to a military operation. Pools should be as large as possible and disbanded at the earliest opportunity--within 24 to 36 hours when possible. The arrival of early access pools will not cancel the principle of independent coverage for journalists already in the area.
C. Even under conditions of open coverage, pools may be appropriate for specific events, such as those at extremely remote locations or where space is limited.
D. Journalists in a combat zone will be credentialed by the U.S. military and will be required to abide by a clear set of military security ground rules that protect U.S. forces and their operations. Violation of the ground rules can result in suspension of credentials and expulsion from the combat zone of the journalists involved. News organizations will make their best efforts to assign experienced journalists to combat operations and then make them familiar with U.S. military operations.
E. Journalists will be provided access to all major military units. Special operations restriction may limit access in some cases.
F. Military public affairs officers should act as liaisons but should not interfere with the reporting process.
G. Under conditions of open coverage, field commanders will permit journalists to ride on military vehicles and aircraft whenever feasible. The military will be responsible for the transportation of pools.
H. Consistent with its capabilities, the military will supply PAOs with facilities to enable timely, secure compatible transmission of pool material and will make these facilities available whenever possible for filing independent coverage. In cases when government facilities are unavailable, journalists will, as always, file by any other means available. The military will not ban communications systems operated by news media organizations, but electromagnetic operational security in battlefield situations may require limited restrictions on the use of such systems.
I. These principles will apply as well to the operations of the standing DoD National Media Pool system.

Appendix C

PUBLIC AFFAIRS ESTIMATE

SAMPLE PA ESTIMATE

FM 6-99 (101-5)

____

Classification

Headquarters
Place of issue
Date, time, and time zone
Message Ref. no.

PUBLIC AFFAIRS OPERATIONS ESTIMATE NO. __________

References: Maps, charts, or other documents.

Time zone used throughout the estimate:

1. MISSION

This paragraph lists the command's restated mission from a public affairs perspective.

2. THE SITUATION AND CONSIDERATIONS

The paragraph describes the strategic and operational media environment in which the operation is being conducted and identifies the critical factors that might impact on the command's mission -- the "action and reaction" within global media channels. It identifies the media environment across the operational continuum, describing it from "austere" for low media interest and capability in a limited AOR communication infrastructure to "dynamic" for high media interest and capability in a high-tech AOR infrastructure. At a minimum, this paragraph must include:

a. Information environment. This paragraph describes the characteristics of the operation and the information environment in the area of operations. It identifies any activities or issues affecting the over-all mission and the command's public affairs objectives.

b. Media presence. An assessment of the news media presence in the theater of operations prior to deployment and the likely presence of additional news media during the conduct of operations. This assessment should address the authority under which media representatives are operating and the degree of control that can be imposed on their efforts.

c. Media capabilities. An assessment of the media's information collection and communication technology, specifically identifying their level of visual information acquisition and satellite communication capabilities. It includes an analysis of the logistics support, transportation assets, and host-nation communications infrastructure available to them.

d. Media content. An assessment of the global media's presentation of information and their agendas, and an analysis and prioritization of the potential strategic and operational issues confronting the command in the news media. This media content analysis will provide an evaluation of the quantity of coverage and the nature of that coverage.

e. Public opinion. Assessment of national and international attitude about the operation and command, leaders, and soldiers conducting it. This paragraph should include both the perceptions held by major audience groups and the relative solidity or strength of those attitudes. A public opinion analysis should include as a minimum an analysis of the following groups:

- American public
- Civilian political leaders
- Coalition and allied forces and their publics
- International audience
- Internal command audience
- Home station public

f. Information channel availability. An assessment of the information channels available for the communication of information in and out of the AOR. It identifies the means available to the commander for receipt, transmission, and dissemination of voice, data, text, graphics, and digital visual imaging. It describes command, coalition, and local national facilities and equipment available, to include an analysis of available telephone lines for voice and data transmission, the accessibility of audio and video channels, the prevalence of private communications devices such as soldier-owned cellular telephones, facsimile machines, computers, portable radios and televisions, still and video cameras, and the nature and flow of the information possible through these channels.

g. Information needs. This is an assessment of the information needs of the previously identified key publics. It analyzes key internal and external audiences and assesses theirs news and information expectations. It identifies the types of information made available to these key audiences.

h. Personnel situation. Describes the present dispositions of public affairs personnel and units that affect the PA mission, and the assets needed and available. State known personnel problems, if any, that may affect the PA situation. Consult the personnel estimate for details. (Examples of personnel include shortages of PA NCOs and skilled operators for equipment.)

i. Public Affairs situation. This summarizes current PA objectives and identifies specific courses of action for each objective. At high levels of command, detail information in a summary with a reference to an annex to the estimate. Subparagraphs will include all current (PAG) from OSD (PA).

j. Logistical situation. State known logistic problems, if any that may affect the PA situation. See logistic estimate for details. (Examples of logistic problems include the lack of transportation and adequate facilities.)

k. Assumptions. Until specific planning guidance from the commander becomes available, you may need assumptions for initiating planning or preparing the estimate. Modify these assumptions as factual data or planning guidance becomes available.

3. ANALYSIS OF COURSES OF ACTION

Analyze each course of action based on the public affairs objectives in paragraph 2i. Indicate problems and deficiencies. At a minimum, subparagraphs should include media facilitation and support, news and information provision, and force training and support.

Analyze each COA from a PA point of view to determine its advantages and disadvantages for conducting PA. The detail in which the analysis is made is determined by the level of command, scope of operations, and of urgency of need.

4. COMPARISON OF COURSES OF ACTION

Compare each course of action. List advantages and disadvantages of each course of action under consideration. Include methods of overcoming deficiencies or modification required for each course of action.

5. CONCLUSIONS

a. Indicate whether you can support the command mission (restated in paragraph from the public affairs viewpoint.

b. Indicate which COAs you can best support.

c. List major public affairs deficiencies, which commander must consider. Include specific recommendations about methods of eliminating or reducing their effects.

____

(Public Affairs Officer)
ANNEXES: (As required)

DISTRIBUTION:

(NOTE: The headings listed in this assessment are for example only. Use headings appropriate to your command's operations.)

____

Classification

PA Operations Estimate

Appendix D

PUBLIC AFFAIRS ANNEX

SAMPLE PA ANNEX

FM 6-99 (101-5)

____
(Classification)

(Change from oral orders, if any)

Copy ___ of ___ copies
Issuing headquarters
Place of issue (may be in code)
Date-time group of signature
Message reference no. _____

ANNEX__ (PUBLIC AFFAIRS) TO OPERATIONAL ORDER ______

References: Maps, charts, and other relevant documents

Time zone used throughout order:

1. SITUATION

A brief general description of the situation, information affecting public affairs support, which paragraph 1 of the OPORD does not cover, and intended purpose of this annex.

a. Friendly forces. Outline the higher headquarters' plan (and PA annex) and adjacent unit PA plans. Provide information on friendly coalition forces, which may impact the PA mission. Note PA resources supporting the unit. (Who, where, when), (higher, allied and adjacent headquarters).

b. Attachments and detachments. Identify all augmenting PA units supporting this command and all attached/assigned subordinate units. Include effective dates, if applicable.

c. Enemy forces. List information not included in the OPLAN/OPORD, which may impact the PA mission. (Who, where, when, disinformation, rumors, propaganda and OPSEC).

d. Media. Identify media in the area. (who, where, pools, US. international, local-host country).

e. Assumptions. List any additional assumptions or information not included in the general situation, which will impact the PA mission.

2. MISSION

Clearly, concisely state the public affairs mission. (Internal information for deployed and non-deployed forces, media facilitation and staff operations).

3. EXECUTION

a. Concept of operation. Briefly summarize the public affairs operation plan. Include PA priorities. (Intent --access, information, welfare, morale, will to win) (Concept--who, where, what, why, when) (Specifics--task to a subordinate; who is to do what, where, when, covers non-PAs too, actions with media: credential, train, transport)

b. Outline of PA tasks. Identify and assign supporting PA tasks to each element of subordinate and supporting units. Assign specific tasks to elements of the command charged with public affairs tasks, such as requirements for PA augmentation.

c. Coordinating instructions. Give details on coordination, task organization and groupings. List instructions, which apply to two or more subordinate elements or units. Refer to supporting appendixes (PA assessment) not referenced elsewhere. (Public Affairs Guidance, media in country, media enroute with US forces, media contact report, handover checklist, and task organization).

4. SERVICE SUPPORT

a. A statement of the administrative arrangements applicable to this operation. If they are lengthy or are not ready for inclusion in the OPORD, these arrangements may be issued separately and referenced here.

b. A statement of the logistical arrangements applicable to this operation. Specific coordination should be included if possible, but arrangements may be issued separately and referenced here, if they are too lengthy. (Class I-IX and water), (Services: billets, medical, laundry and mortuary), (Transport: ground, air, TOE, tasked rented/leased, contracted).

5. COMMAND AND SIGNAL

List signal, visual imaging and satellite communications policies, headquarters and media center locations or movements, code words, code names, and liaison elements. (PAO location, media center, JIBs, sub-JIBs, phones, faxes, e-mail and web page).

ACKNOWLEDGE:

NAME (Commander's last name)
RANK (Commander's rank)

OFFICIAL:
APPENDIXES: (List PA assessment appendix)
DISTRIBUTION:

Classification

Appendix E

PUBLIC AFFAIRS GUIDANCE

SAMPLE PA GUIDANCE

Public Affairs Guidance (PAG) is the operational tool that guides commanders and their public affairs officers in the application of doctrine and policy during major military operations, exercises, and contingencies. The information below is tended to assist local commanders in preparing and obtaining approved guidance.

DoD policy requires that PPAG be provided to the Assistant Secretary of Defense-Public Affairs (ASD-PA) by the Unified and Specified commands and others, as required for all major operations.

This requirement includes major training exercises that could attract national and/or international attention. PPAG may not be used without ASD-PA approval.

Upon receipt of the warning order, the commander, through his PAO, should request PAG from high headquarters. PAG may be included in alert notification or operational orders (see Part Three: Operational Planning). Commanders of major units/commands will direct their PAOs to prepare PPAG to forward the proposal through MACOM and Unified/Specified command PA channels to ASD-PA.

Commanders of Unified/Specified commands should ensure that the PPAG has been coordinated with appropriate organizations within the theater of operations whenever possible (e.g., embassies, country teams, host governments, subordinate commands).

Upon receipt of the PPAG, the ASD-PA coordinates and staffs the PPAG within the DoD and Department of State.

The ASD-PA then issues a message either approving, modifying, or disapproving the PPAG. PPAG is broken down into subject, references and then eight paragraphs (Information/explanation, Purpose and coordination for PPAG, PA Approach, Public Statement, Q&As, Contingency Statement, Miscellaneous Information, Point of Contact). The format for PPAG follows:

SUBJECT

The subject line of the PPAG should state "PROPOSED PUBLIC AFFAIRS GUIDANCE - followed by the exercise and/or event name (U)." For coordination, it is best if the subject is unclassified.

If an exercise or event is so sensitive that the actual name cannot be used, an unclassified short title should be used; e.g., "PROPOSED PUBLIC AFFAIRS GUIDANCE - CC-(U)."

REFERENCES

Pertinent messages or other documents shall be cited in the reference section. If the PPAG is based on PA policy in the Significant Military

Exercise Brief, then the SMEB message Date-Time-Group (DTG) shall be listed.

EXERCISE INFORMATION

The **first paragraph** of the PPAG shall explain the references, the exercise, and any significant existing or anticipated problems associated with the exercise. The information in this paragraph is not for release so may remain classified after the PAG is approved for release. This paragraph may restate some PA information from the SMEB.

COORDINATION INFORMATION

The **second paragraph** shall explain the purpose of the message; identify it as being fully coordinated and theater-approved; request ASD-PA approval and specify the date it is required for use. If the PAG is transmitted to the ASD-PA before it is fully coordinated, it is the responsibility of the submitting command to ensure that the ASD-PA is promptly informed of the results of the remaining coordination. The submitting command should always follow-up a PPAG message with a phone call to ensure that the primary addressee(s) is aware that the message is en route. When the submitting command is a supporting CINC from outside the supported CINC's AOR, the supported CINC is responsible for theater coordination.

PA APPROACH

The **third paragraph** shall discuss the public affairs approach for the exercise; i.e., active or passive. This may be a restatement of the PA policy indicated in the SMEB.

- Active Approach
  - For this discussion, an "active approach" involves efforts made to stimulate public or press interest such as distributing press releases and inviting the press to observe the exercise. If an exercise or event is to be publicly announced, this paragraph shall state who will make the announcement, the method of announcement, and preferred time, and date for the announcement. If unusual circumstances prevail, the rationale for the recommendation should also be included. Part I and II exercises shall normally be announced by the ASD(PA) by issuing a news release (blue top). Other lesser exercises or training deployments, if announced by the ASD(PA), normally shall be made by release of a memorandum for correspondents (MFC). The preferred release time and/or date of exercise announcements is 1200 Eastern Standard Time (E.S.T.) on either Tuesdays or Thursdays in conjunction with the normally scheduled DoD press briefing. If a combined announcement is desired with a host country, complete details of the methods, time, and procedure shall be included in this paragraph. The active approach is recommended whenever possible to ensure appropriate media coverage of specific commands and/or units.
- Passive Approach
  - A "passive approach" is where no action is taken to generate media and/or public interest in an issue or activity beyond answering specific inquiries. If a passive approach is desired,

the PPAG shall so indicate and specify that the PAG is for response to query (RTQ) only. It shall also specify who is authorized to respond; e.g., "Only OASD(PA) may RTQ," or "All of the following addressees may use this PAG for RTQ only." To de-emphasize an event, it is best to authorize release or RTQ at the lowest possible level.

PUBLIC STATEMENT

The **fourth paragraph** shall contain a statement that explains the exercise and/or event. The statement shall be for public release in an active PA approach or for RTQ in a passive PA approach. For ease of coordination, each paragraph of the statement shall be identified as a sub-paragraph of the message; for example: The following statement is for initial public release: (TEXT FOLLOWS): QUOTE.

EXERCISE (NAME).....

THE EXERCISE WILL.....

PREVIOUS MILITARY.....

FOR ADDITIONAL INFORMATION,

CONTACT.....(UNQUOTE).

As indicated above, the last paragraph of the statement shall identify points of contact where additional information may be obtained.

QUESTIONS AND ANSWERS (Q&AS)

The **fifth paragraph** shall contain a list of proposed Q&As to enable the user to respond to the majority of anticipated questions. They should all be contained in one paragraph and should be numbered sequentially; e.g., Q1, A1; Q2, A2; Q3, A3, etc. Q&As are for use in both active and passive PA approaches, but are strictly for RTQ only and shall not be given to media as handouts.

CONTINGENCY STATEMENT

The **sixth paragraph** of the PPAG shall contain a contingency statement to be used efore release of the final PAG. Usually, the contingency statement should be that we don't discuss exercises before they have been formally announced. However, this approach can be modified, as appropriate, depending on the circumstances of the exercise. If a contingency statement is not required, so state in Paragraph 6 of the PAG.

MISCELLANEOUS INFORMATION

The **seventh paragraph** shall contain other pertinent information to include the following items (when a certain sub-paragraph is not applicable, so state):Media Information Centers (e.g., Joint Information Bureaus (JIBs), Press Information Centers (PICs), exercise PA elements, etc.) Discuss whether centers are joint or combined; delineate who is responsible for the establishment; give generic description of its composition (e.g., U.S. Army desk (O-4 and E-6/E-7), U.S. Navy/Marine desk (USN O-4/O-5, and USMC E-5/ E-6), etc.); establish the center's functions (coordination of all exercise media and/or PA activities, clearance of U.S. military-generated news material before release, production of news material for release, escort of accredited news media representatives); etc.

- Command Relationships
  - Designation of sole approving authority for all exercise-related news materials; procedures for the release and/or clearance of information (to include list of addressees for notification in case of accident and/or incident); request for participating commands and/or units to ensure that the media center is action and/or information addressee on all messages with potential PA impact (to include incident and/or accident reports); hometown news release requirements and/or instructions (passive PA approach may make hometown releases inappropriate); etc.
- Media Coverage
  - State whether media coverage is encouraged or solicited, giving rationale; news media transportation instructions; point of contact (POC) and procedures for handling such requests; requirements for news media representatives (valid passport, working media visa, local accreditation requirements, funds for food, lodging, return travel (if military air is not available), etc.); instructions regarding assistance to continental United States (CONUS)-based units for handling request from news media for accompanying travel before and following public announcement of the exercise; etc.
- DoD National Media Pool
  - Each exercise is a potential opportunity for activation and deployment of the DoD National Media Pool to cover exercise activities. As a minimum, planning should include arrangements for local ground and/or air transportation, special clothing or equipment to be provided, messing, billeting, protection of media equipment and gear, local escort requirements, and communications support for filing of pool products. Sponsoring commands shall indicate whether the exercise should be considered for a pool deployment. Identify the primary POC should the pool be activated.
- Internal Media and Audiovisual Coverage
  - Provide instructions on assistance that will be provided to this effort; degree of freedom of movement (to include whether escorts are necessary); screening of visual information (VI) materials upon completion of exercise; sponsoring command POC for handling internal information matters; etc. Also include guidelines for Armed Forces audiovisual teams documenting the exercise.
- Media Opportunities
  - If known well enough in advance, provide chronology of potential exercise events that would be of interest to media.
- Miscellaneous PA Considerations

- Indicate any other proposed PA activities or considerations; if there are none, then so state.

POINTS OF CONTACT

The **eighth paragraph** shall state the originating POC's name and phone number.

DECLASSIFICATION INSTRUCTIONS

Declassification instructions shall be the last part of the message and in accordance with subsection 4-207 of DoD 5200.1-R (reference (d)).

Appendix F

PUBLIC AFFAIRS OPERATIONS BRIEFINGS

BRIEFINGS

The purpose of a briefing and the desired response or result determines the briefing technique. Basically, there are four briefing types: the information briefing, the decision briefing, the mission briefing, and the staff briefing.

The Information Briefing. The information briefing informs the listener and deals primarily with facts and background information. The Information Briefing contains an introduction to the subject and the scope of the subject area. It then presents the high-priority information requiring immediate attention and complex information involving complicated plans, systems, statistics, or charts. It may also explain controversial situations or information, which require elaboration.

The Decision Briefing. The Decision Briefing includes many of the elements of the information, but goes further by seeking a decision from the decision-maker. At the beginning, the briefer clearly states that he is seeking a decision; at the end, he requests the decision.

The Mission Briefing. The Mission Briefing gives special instructions, amplifies the mission, elaborates on new orders, or assigns taskings to subordinate elements. This briefing usually follows the five-paragraph operations order format. But the briefer may also choose the information briefing format.

The Staff Briefing. The Staff Briefing informs the commander and staff of the current operational situation. Its purpose is to generate a coordinated or unified effort and in a tactical environment. It serves to keep the entire staff aware of each section's activities, thus aiding coordinated action. While there is no specific briefing format, commanders usually tailor this type of briefing to fit their information needs. PAOs address the major PA activities and the PA implications of the operational situation and other staff sections activities.

In a commander's staff briefing, the PAO is responsible for providing a summary of the "global information environment (GIE)." The PAO should present this information at the beginning of the briefing, following the intelligence summary. The PAO's GIE summary, combined with the G2/S2's intelligence preparation of the battlefield (IPB) briefing, helps to complete the picture of the operational situation, which the other elements of the command must react.

Preparation. After conducting the initial analysis of the situation, the briefer gathers and organizes information, prepares visual aids, and then rehearses, briefing only what is essential. Before developing the briefing, the briefing officer or NCO must know and understand:

- the higher headquarters mission
- the higher commander's intent
- the commander's personal guidance and intent
- all aspects of the PA estimate of the situation
- the command's objectives

While much of the information required for a GIE presentation can be found in previously published PA Area Studies, a significant amount of situational information should pulled directly from the PA Estimate of the Situation and updated to present the latest information available.

BRIEFING FORMAT

The PA briefing format should follow an outline similar to the PA Estimate of the Situation:

- Mission - Command mission including PA objectives
- Situation and Conditions
- Information environment
- PA situation
- Media presence
- Media capabilities
- Media content
- Public opinion
- Information channel availability
- Information needs
- Current Operations/Actions
- Future Operations/Actions

Appendix G

PUBLIC AFFAIRS AREA STUDIES

FORMAT FOR PA AREA STUDIES

Public Affairs area studies provide the PA specialist and the PA planner a starting point to begin their campaign planning process. These studies are produced either to address long-term general background information or to address immediate short-term needs.

BASIC STUDY

A PA study is a document that succinctly describes the most PA-pertinent characteristics of a country, geographical area, or region. It serves as an immediate reference for the planning and conduct of PA operations.

TITLE PAGE

The title page of a Media Environment Study (MES) must show the country, area, group, or other subject of the study. It should also show the classification, the copy number, the date-time group, and command post location. (The originating unit should keep a record of the names, grades, and duty assignments of the authors.)

EXECUTIVE SUMMARY

The executive summary should address the strategic significance of the area under study. (The authors should write the executive summary last, in a clear, brief, accurate, and coherent form.)

NOTE: The commander reading the summary should us its information to decide how to employ Public Affairs in that area.

TABLE OF CONTENTS

Executive Summary

Introduction

Chapters

1. History and Ideology
2. Government and Politics
3. Foreign Relations and Policy

INTRODUCTION

The introduction outlines the study's intent, cites the directive requiring the study, and explains the study's format.

The study's intent is to provide a summary of aspects of subject country significant to GIE. The study should identify psychological vulnerabilities, characteristics, insights, and opportunities that exist in subject country.

It is prepared as a basic source document for further development of estimates, plans, and annexes.

Although this study can help develop contingency plans, it is not tailored to any particular plan. Rather, its neutral data and insights can be used to analyze possible political and military developments in the region.

Insert here a paragraph referencing the authority directing the study and stating the research cutoff date and provisions for updating the study.

Focus the PA study on the GIE aspects of the many topics addressed. Do not view this document as a comprehensive and self-contained area study.

Instead, use it as a complement to such other standard references as the Department of the Army Area Handbook Series. In addition, the PA study should include:

Results from standard open source products.

Tries to be more analytical than descriptive in nature, making it subject to varying individual perspectives.

Works well when PA specialists compare it with studies on other countries in the region or area.

Is not a U.S. foreign policy statement or comprehensive analysis of subject country, except in areas with direct GIE relevance.

Insert here a statement of U.S. policy goals toward the country in question. This information comes from the proper USA Country Plan, Department of State Policy Memorandum, or similar document, in the priority order.

Because certain gaps exist in getting PA study material (classification level, availability of complete and timely information, or time limits on research), listing these gaps here to aid future research and guide PA study users to further inquiries.

Insert here issues or GIE-relevant material (such as an area map) not included, addressed, or completely answered elsewhere.

CHAPTER 1

HISTORY AND IDEOLOGY

In this chapter, review the evolution of the state and its people, focusing on aspects having GIE and Public Affairs significance. Do not detail chronology of the country's development. Keep in mind, however, the country's history has an important relationship to the country's historical perspective, attitudes, and current world position.

Because of its special importance to PA, in this chapter, cover historical issues thoroughly. An historical analysis of current political, economic, and military policies gives PA personnel a solid base for the study.

CHAPTER 2

GOVERNMENT AND POLITICS

In this chapter, summarize the country's political system. Give a description of its political power sources, policy making process, and the political complexities of the government.

When discussing the political system, pay close attention to the role of individuals, special interest groups, and political parties. Include the population's political attitudes, values, and view of the political system. Also discuss the government's function in society.

CHAPTER 3

FOREIGN RELATIONS AND POLICY

In this chapter, summarize the country's foreign relations. Describe its political alignment in world affairs and its relationship with the United States. Describe the foreign policy of the country. Also analyze and interpret why the country acts as it does in international affairs.

CHAPTER 4

SOCIETY AND CULTURE

In this chapter, analyze the subject country's social setting. Provide the PA personnel with the knowledge needed to understand potential audiences.

Cover the country's social organization and cultural and behavioral patterns and characteristics. Place special emphasis on the society's social values and the role of the family.

Address culture, social organization, education, customs, ethnic composition, and the interrelated effects of religion, language, and history.

CHAPTER 5

ECONOMY

In this chapter, present a brief description of the characteristics, structure, and dynamics of the national economy. Cover the subject country's economic strengths and weaknesses, current economic and labor problems, and economic potential

Describe the country's economic base and the importance of agriculture, industry, and trade. This information helps determine if the present economic structure meets the people's needs.

These economic considerations explain many of the sociological conditions that impact public opinion. Address society's perceptions of the wisdom of government economic policies.

Also describe individual or group perceptions of how members of society stand to gain or lose from those policies.

CHAPTER 6

MILITARY ESTABLISHMENT

In most countries, the military establishment involves itself in internal politics as well as external defense. Even when the military establishment does not directly compete for political power, its actions influence social and political development. Analyze the following topic areas:

- Emergency of the modern military establishment.
- Military roles in the political, economic, and social spheres, and the effects of those rules.
- Issues creating cohesion or conflict within the armed forces.

Leadership.

Conflicts within the military establishment.

Extent, quality, and influence of foreign military aid.

CHAPTER 7

COMMUNICATION PROCESS AND EFFECTS

List essential information about communication patterns for the implementation of a PA program. Include the following information:

Manner and social means of communication (not technical data on communications facilities).

Languages and language groups, nonverbal communication, and nonverbal symbols specific to the country's culture or cultures.

Distinctive styles in rhetoric or visual arts, including dramatic, poetic, and musical forms. All these forms are significant to PA.

Data on the society's formal and informal leadership positions where the incumbents are key communicators and opinion leaders.

Analysis of the reading and listening habits of the society.

Analysis of printed formats.

Analysis of media effectiveness. Also address freedom of the press issues, if any.

ENDNOTES

LIST PUBLICATIONS, DOCUMENTS, AND OTHER SOURCES USED IN PRODUCING THIS BPS. NUMBER TEXT REFERENCES SEQUENTIALLY WITHIN THE STUDY.

APPENDIX A

COUNTRY SUMMARY

Give the reader a brief overview of the subject country, its geography, and its people. These background data and statistics should include the following items:

Country. Identify the country, tell when formed, and show previous control.

Government. State briefly the type of government, method of appointing or electing leaders, and length of terms. List current leader or leaders and political power in country.

Size, location, and geographical subdivision. List the size of the area in square miles or kilometers, and give the general location. Show any geographical subdivision, such as coasts, mountains, and flatlands.

Population. State the number of people and the area density. Show the heavily populated areas.

Languages and dialects. List the official language, languages spoken by the population, percentage of population speaking each language, and areas of the concentrations.

Labor. Outline the total work force, the area of endeavor, and the percentages.

Religions and sects. List the religions of the area and the percentage of the population that practices each.

Education. List the types of systems and the primary emphasis of each.

Literacy rates. Latest statistics.

Health. List the general conditions of the populace. Describe the medical care system.

Justice. Describe the justice and court systems.

Administration. Outline the breakdown of the governmental and judicial districts, counties, or precincts.

Transportation. List the methods of transportation available and include the total capabilities. This information may include the number of airlines, airfields, kilometers and kinds of highways, and kilometers of waterways and depth.

Armed forces. List organization and strengths.

Police. List the types and areas of responsibilities.

APPENDIX B

GOVERNMENT STRUCTURE

List the formal government structure, key positions, and organizations of the country. Outline the chain of government control, including political parties, if applicable. A schematic diagram may be helpful.

Include only branches of government and their key positions, not names.

APPENDIX C

COMMUNICATION FACILITIES

Give a brief overview of the subject country's media facilities. Include the facilities' locations and levels of technical sophistication. Cover printing, publishing, and the distribution of radio and television receivers, studios, transmitters, and relay facilities. Include news service facilities. Write this appendix as if the U.S. PA units will use this equipment or contract for its services. Since PA personnel may get operating supplies or repair parts from in-country sources, provide the following information:

- Make.
- Model.
- Type.
- Series.
- Name of manufacturer.

Any other technical information on the repair or operation of this equipment.

GLOSSARY

Prepare a glossary that lists in alphabetical order all acronyms and foreign words used in the study. List also all words and terms that have special meaning and need to be defined.

BIBLIOGRAPHY

List the source material used. Include the name of the author, the title of the publication, the publisher, and the date of publication.

DISTRIBUTION

Dissemination is accomplished by the originating agency for the recipients within PA. Include in the distribution list the identification of recipient agency (by code), the number of copies furnished, and the office symbol of the recipient.

Appendix H

BRIEFINGS AND PRESS CONFERENCE FORMATS

NEWS BRIEFING AND PRESS CONFERENCE FORMAT

Before the Presentation

Know your publics
Anticipate interests, concerns and questions
Consider the latter in preparation

Prepare your presentation
Develop a strong introduction
Develop a maximum of three key messages
Assemble your supporting data
Prepare audiovisual aids
Practice

Prepare for answering questions
Anticipate what questions will arise
Prepare answers to those questions
Practice questioning and responding

The Opening Statement

A strong opening statement sets the tone for the press conference or news briefing and is crucial in attempting to establish trust and build credibility. The elements of a strong opening are:

Introduction
A statement of personal concern
A statement of organizational commitment and intent
A statement of purpose and plan for the meeting

Key messages and supporting data
A maximum of three "take-home points"
Information to support the key messages

Conclusion
A summarizing statement

Total time for all presenters should be 15 minutes or less. Do not have too many presenters. Three is usually sufficient.

Introduction

Remember that perceived empathy is a vital factor in establishing trust and building credibility and your publics assess it in the first 30 seconds.

Examples are:

Statement of personal concern: "As a resident of this community I'm interested in the safety and well-being of our families and neighborhoods."

Statement of organizational commitment and intent: " I'm here to share with you the knowledge and confidence I have in the military's ability to assist the citizens of our community. They have been trained in their occupational skill to assist with the task at hand."

Statement of purpose and plan for the presentation: Today I would like to share with you the most current information regarding the (incident.) I will also be available to answer additional questions or to continue the discussion.

Key messages and supporting data

The key messages are points you want your publics to have in their minds after the presentation. They should:

Address central issues.
Be short and concise.
Examples are:
"We have trained personnel and emergency response plans in place to aid in protecting the health, safety and welfare of the public. We are working with local and state officials to handle the incident.

"We are actively responding to the emergency...."

To develop your key messages:

Brainstorm
Think freely and jot down all pieces of information you wish to communicate.

Select key messages
Identify the most important ideas. Repeat the process until you list is down to three items.

Identify supporting data
Other information you listed probably provides support to your key messages. Organize it to reflect this.

Conclusion

Restate verbatim your key messages.

Add a future action statement --- What is your organization going to do about this problem in the short and long term?

Appendix I

WAIVER OF LIABILITY STATEMENT

Whereby, I <u>NAME</u>______________________ passport no:____________ am about to travel with __________forces, and whereas I am doing so entirely upon my own initiative, risk and responsibility; now, therefore, in consideration of the permission extended to me; I do hereby for myself, my heirs, executors and administrators, remiss, release and forever discharge ________ and its member officers, agents and employees acting officially or otherwise, from any and all claims, demands actions or causes of action, on account of my death or on account of any injury to me or my property which may occur from any cause during my stay, travel as well as all ground, flight or sea operations incidents thereto.

I also agree to withhold any classified information, which may be accidentally disclosed to me, and to respect embargo restrictions, which may be imposed on information which, if disclosed, may jeopardize operational security. During my stay with _______ forces, I will not interfere with operations. I understand that failure to comply with these security restrictions will result in the loss of authorization to accompany ______and may result in cancellation of my press registration.

____________________	____________________
Signature	**Witness**
____________________	____________________
Printed Name	**Nationality**

Address:____

Please provide the following information for a person to be notified in an emergency (preferably next immediate relative):

____

____

Appendix J

SAMPLE PRE-DEPLOYMENT CHECKLIST

INTRODUCTION

A multitude of factors make each deployment uniquely different from other deployments. Each factor must be carefully examined to determine its impact on the mission, actions before deployment and what equipment is taken.

(Unit Basic Load). In all other instances the chances of drawing the UBL are remote. Certainly if the unit is deploying for war it would draw the unit's basic load of ammunition Exercise or deployment duration will affect the quantity of expendable supplies.

This checklist, like all others, is based on what has occurred before and what we have come to expect in the future. As equipment and missions change, so too must the checklist. Bottom line, expect the unexpected and react accordingly; use the checklist as a guide to help you through deployment preparation. Add to it whenever the need arises.

CHECKLIST ITEMS

1. Each HQ element and unit should have a 45-day supply of expendables identified and set aside for contingency operations. This supply should be inventoried quarterly and stockage rotated accordingly.
2. Maintain and update a master list of all areas and topics that can affect individual readiness. This includes all shots, dental status, family support plan, check to bank, etc.
3. The PAO or detachment commander (or deploying team leader) should:
   a. Review and update PA estimates annexes and plans.
   b. Obtain and comply with applicable published Public Affairs Guidance.
   c. Inventory and inspect TOE equipment for accountability and serviceability before deployment.
   d. Figure total weight and cube of equipment before moving to assembly area.
   e. Check dependency clause in TOE document or OPLAN to determine who will be supporting and who is supporting us. Ensure responsible parties know the relationship. Establish liaison with units OPCON; attached or any new parent organization.
   f. Ensure accompanying equipment not on TOE is listed on interim authorization document (such as facsimile machines or cellular phones).
   g. Prepare an internal OPORD for your element detachment or team.
   h. Review the essential elements of friendly information (EEFI) contained in the base OPLAN/OPORD and ensure each soldier understands them.
4. All leaders must consider the need for the following:
   a. Will flak vests be required?
   b. Should each soldier take a footlocker instead of a dufflebag?
   c. Will weapons' carrying/security cases be needed/available? (Will weapons and protective masks be required?)
   d. Will desert or jungle uniforms be required, and if so, what fund can be used to pay for alterations, sewing and patches?
   e. Is there an opportunity for a service contract to be initiated at the deployed location? Cameras, video, computers, etc.? if not, and cameras go down and must be swapped out, what is the plan?
   f. Will the deployment be considered TDY? If so, who prepares orders? What fund cite will be used? Are rental vehicles available? (Are we TDY under field conditions and required to carry meal card?)

g. Will a fund cite be made available after redeployment to pay for repairs? (Make this request soon after being tasked - don't wait until redeployment to find out you've got to use unit funds to repair equipment.)
h. If departing from location other than current station, what type of transportation to that location is to be used for personnel and equipment?
i. Will personnel and equipment travel together at all times? (Insist that they do whenever possible!)
j. Does each team have a credit card holder for authorized payments or purchases?
k. Has an express mail system been established to ensure timely transit of products to home station? With what frequency? (If you are the ARFOR or sub-JIB, have the division PAO's/PAD's establish a plan to get products to you for review, release or use).
l. Under what conditions will the MPAD or team(s) work? Fixed site, field conditions, etc. Will we need to deploy our own tents for sleep and work?
m. Is the heat a condition that will affect computers, cameras, batteries, etc? If so, is air conditioning available? Refrigerators for film? What about humidity problems? Air conditioning may help but can dehumidifiers do better...consider charcoal bags placed in shipping cases.
n. From what unit(s) do we draw support...rations, billets, fuel, etc.?
o. If one team is deployed for a lengthy duration, is there an opportunity for the teams to be rotated?
p. If deploying a risograph or other commercial printing machine, what quantity of copies and frequency (daily, weekly) is desired? What is the plan for distribution of copies? What is the plan for paper replenishment?
q. Has every effort been made to ensure deployed asset has commercial phones available to assist transmitting digital photographs, responding to query, accidents and incidents, fax capability, etc.
r. Has unit/team packing list been carefully checked to ensure the easy-to-forget yet must- have items are not forgotten. Glue sticks, chalk, scotch tape, blank overheads, etc.

Appendix K

PAO SOP OUTLINE

Chapter 1, Alert Notification Procedures
- Annex A, Notification Procedures
- Annex B, Section Telephone Contact Roster

Chapter 2, Individual Preparation
- Annex A, Individual Admin Checklist
- Annex B, POV Storage Plan, Procedures for Completing Post POV Storage Forms.
- Annex C, Personal Property Storage Plan
  - Section 1, Power of Attorney
  - Section 2, State of Obligations
  - Section 3, DD Form 1299, Application for Shipment/Storage of Personal Property.
  - Section 4, DD Form 1701, Household Goods Inventory
- Annex D, Recommended Personal Readiness Equipment
- Annex E, Family Member Pre-deployment Checklist

Chapter 3, Unit Preparation
- Annex A, Unit Equipment List

Chapter 4, Tactical Vehicle Preparation
- Annex A, Vehicle Preparation Standards
- Annex B, Vehicle Load Card
- Annex C, Vehicle Inspection Checklist
- Annex D, Joint Airlift Inspection Record
- Annex E, HMMWV Configuration and Load Plan

Chapter 5, Rear Detachment Operations
- Annex A, Communications
- Annex B, Logistics Coordination

Chapter 6, Public Affairs Checklists
- Annex A, Pre-Deployment Checklists
- Annex B, Guide for Media Interviews
- Annex C, METL and Tasks, Conditions, Standards
- Annex D, DoD Principles of Information
- Annex E, Policy on the News Coverage of U.S. Military in Combat
- Annex F, PA Guidance on Terrorism Counteraction
- Annex G, PA Guidance on Counter-Drug Operations
- Annex H, Standard PA Ground Rules
- Annex I, Spokesperson Guidelines
- Annex J, Command Unique Media Operations Center Guidelines

Chapter 7, Field Operations
- Annex A, Tactical Uniform
- Annex B, Installation Security
- Annex C, Tactical communication
- Annex D, Personal Hygiene
- Annex E, Morale, Welfare and Recreation

Appendix L

EXAMPLE OF MEDIA OPERATIONS CENTER

Operations of a media center will need the following support:

- Communications
- Vehicle support (day-to-day operations and media transport -- tactical or non-tactical as needed)
- Billeting and rations for media center personnel
- Admin support personnel for 24-hour operations
- Office space (hard site if possible) and power as needed

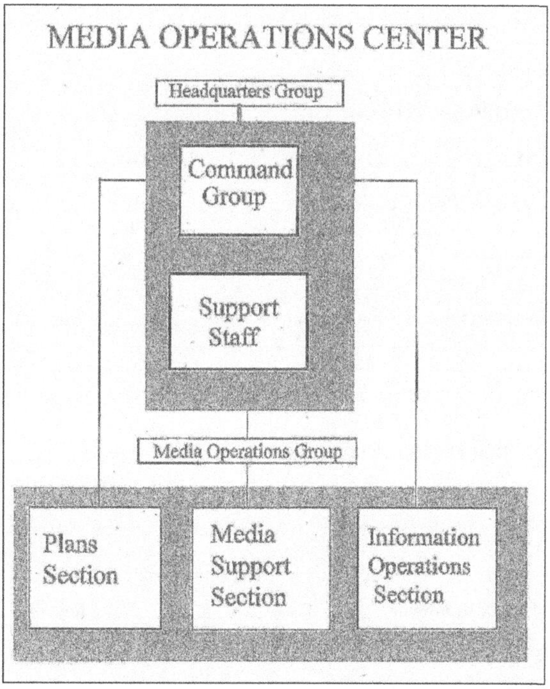

Figure L-1. Media Operations Center

Appendix M

NEW MEDIA INQUIRY FORMAT (SAMPLE)

This format is intended as an example only and should be adapted to local needs and SOPs. In addition to query forms, PAOs should maintain a separate log of all inquiries.

MEDIA QUERY SHEET

Query Number:____________

Deadline:_________________

CALLER'S NAME:____

CALLER'S NEWS ORGANIZATION:____________________________________

CALL TAKEN BY:________________TIME:____________DATE:_____________

QUESTION (use reporter's precise wording):_________________________________

RESPONSE (if written release is made, attach a copy):__________________________

SOURCE OF INFORMATION/COORDINATION (e.g., G-3, surgeon): ____________

RELEASED TO:____

TIME/DATE:____

RELEASE METHOD: In Person________ Phone_____________ News Release_______

RELEASED BY:_______________ RELEASE NUMBER:__________________

Appendix N

GUIDE FOR MEDIA INTERVIEWS

GUIDELINES

1. Considerations

a. When human safety or other serious concerns are involved, deal with those considerations first.

b. Communicate only information that is approved for external distribution. Always tell the truth.

c. Know to whom you are speaking. Get the person's name and telephone number, if necessary.

d. Do not be intimidated. You may tell a reporter that you need to clarify an important matter before you can answer questions.

e. Talk from the public's viewpoint. Avoid jargon. Speak within the audience's frame or reference.

f. If the questions do not lie within the framework of approved statements or within your area of expertise, find the appropriate technical advisor or spokesperson.

g. State the most important fact at the beginning. Place your own headline on the answer.

h. Attack problems in your answers, not people.

i. Do not repeat offensive or negative language. Do not let other people put words in your mouth.

j. Direct questions deserve equally direct and forthright answers.

k. Do not exaggerate the facts. Listen to how your answer "sounds" when spoken.

l. Ignore cameras and microphones. Talk to the reporter.

m. During videotaped interviews, it is all right to stop your statement and start over.

n. Do not say "no comment." Explain why you do not have an immediate answer.

o. Keep your composure, even if a news reporter gets snappy.

p. Be prepared to provide sufficient evidence for statements you make.

q. Be especially alert about photos. You have little control over photos taken off military reservation property, but you have every right to control photos taken on the military reservation.

r. Be aware of your surroundings and follow local OPSEC rules when determining interview location

2. What will be asked?

a. What happened and where? When did this occur

b. Are there injuries or deaths as a result? How many and to whom?

c. What actions is the unit taking to control the situation?

d. Have chemicals or other hazardous substances been released into the environment? What kinds? How much?

e. What types of hazards are presented to people off-site?

f. Have off-site emergency response personnel been notified? Which ones?

g. Are unit operations shut down?

h. Has the site or facility been evacuated?

i. How many people are employed at this site?

j. What do you do at this site?

k. How old is the facility? Does it meet current regulations?

l. Why did this situation occur? (DO NOT SPECULATE.)

m. Are there safety rules covering the situation? Were they violated?

n. Has a Site Emergency Response Plan been activated? What does that involve?

o. Tell me about your organization?

p. Will this situation have national ramifications, or will its effect likely be limited to a single site or region?

q. How much money is this going to cost the taxpayers?

r. Is there insurance coverage for the loss or damage? How much?

s. Are commanders handling the situation locally or is a higher headquarters taking control?

t. Has this occurred anywhere within the unit before? Why weren't you ready?

u. What do your soldiers think about this situation?

v. For accidents and incidents, don't speculate causes. Use "ongoing" investigation statements.

Appendix O

MEDIA RESEARCH AND ANALYSIS

AN INTRODUCTION

Since the early part of the 20th century, when there was almost no interest in the size of audiences or in types of people that comprised various audiences, business leaders and their organizational communicators have increasingly come to rely on research for nearly every major decision they make. This expanded demand for information has created a particular demand for media communication research capabilities, specifically -- the development of a scientific basis for media analysis and media effects.

The importance of mass communications research and media analysis is partly due to the realization that gut feelings or reactions are not entirely reliable or credible bases for decisions. Although common sense is often accurate, Army commanders and other decision-makers need additional, more objective information to evaluate problems, especially when lives are at stake. Thus, the past 50 years have witnessed continuing evolution of media analysis, combining research and intuition to create a higher probability of success.

This evolution has resulted in a "scientific" approach to media research known as media content analysis.

SCIENTIFIC RESEARCH: A DEFINITION

Scientific Research is defined as a **"systematic, controlled, empirical, and critical investigation of hypothetical propositions about the presumed relationships among observed phenomena."** This definition contains the basic terms that are necessary in defining the method of scientific research and describes a procedure that has been accepted for centuries. In the 16th century, for example, scientist Tycho Brahe conducted years of systematic and controlled observation to prove wrong many of Aristotle's theories of the universe. By gaining an understanding of the phenomena, he challenged the accepted beliefs and knowledge of the time with his own hypotheses. Thus, scientific research was begun.

Whether we realize it or not, we all conduct research as a matter of course in our day-to-day life whenever we speculate about the possibility of something -- we start with an idea or concept and test it.

All research begins with a basic question or proposition about a specific phenomenon -- for example, Why do Americans usually support the soldiers within the Army when they may not support the operation the soldiers are involved in? What factors determine why Americans will support the political justification for military involvement? What types of messages are most effective in garnering support for American forces?

The answers to these questions can be forecast to some degree with well-designed research studies. There are some difficulties, however. The Army media analyst faces the problems of determining which data collection methods can most accurately provide answers to the questions at issue, and in gaining adequate access to information prior to and during military operations. In the pages that follow, we will describe the methods and procedures PA professionals may use in overcoming these difficulties.

RESEARCH APPROACHES

There are several research approaches or "methods of knowing" which have been used to conduct studies: intuition, authority, and science.

In the **intuition** approach, one assumes that something is true because it is "self-evident" or "stands to reason." An example of this type of thinking would be if some Public Affairs leaders resist efforts to perform area studies because they believe they already "know" their AO.

The **authority** method seeks to promote a belief in something because a trusted or credible source says it is true. Here, the emphasis is on the source, not the methods the source may have used to gather his information.

The **scientific method** approaches learning as a series of small steps, with each step identifying more specific information and leading to a more clearly identifiable conclusion.

For example, one study or source provides only an *indication* of what may or may not be true; the "truth" is found only through a series of objective analyses.

This means that the scientific method is self-correcting in that changes in thought or theory must be continually reviewed, that issues and situations require constant monitoring.

The scientific method has become a valuable tool to produce accurate and useful data in mass media research. This annex focuses solely on the scientific approach and forms the fundamental basis for media content analysis in Army Public Affairs.

MEDIA RESEARCH: THE SCIENTIFIC METHOD

The goal of Public Affairs media research is to provide the methodology to support situational assessment, planning and decision-making that is fast, inexpensive, reliable and valid. The application of scientific methodologies to media research by Public Affairs personnel accomplishes this goal.

Five basic characteristics, or tenets, distinguish the scientific method from the other methods of research. A research approach that does not follow these tenets cannot be considered a scientific approach:

- **Scientific research is objective.** Science tries to rule out eccentricities of judgment by researchers. When a study is undertaken, explicit rules and procedures are constructed and the researcher is bound by them, letting the chips fall where they may. Objectivity also requires that scientific research deal with facts rather than interpretations of facts.

- **Science is empirical.** Researchers are concerned with a world that is knowable and potentially measurable. (Empiricism derives from the Greek word for "experience.") Analysts must be able to perceive, understand, and classify what they study and reject nonsensical explanations of events. For example, a newspaper editor's claim that declining readership rates are "God's will" would be rejected by scientific researchers because such statements cannot be perceived, classified or measured. Experience shows that there are usually easily identifiable reasons for declining readership.
- **Scientific research is systematic and cumulative.** No single research study stands alone, nor does it rise or fall by itself. Astute research analysts always use previous studies as building blocks for their own work. One of the first steps taken in conducting research is the review of all available literature on the topic so that the current study will draw on the heritage of past research.
- **Scientist attempt to find order and consistency in their findings**. In its basic form, scientific research begins with a single, carefully observed event and progresses ultimately to the formulation of theories and laws. A theory is a set of related propositions that present a systematic view of phenomena by specifying relationships among concepts. Researchers develop theories by searching for patterns of uniformity to explain and describe the information collected.
- **Scientific research is predictive.** Science is concerned with relating the present to the future. In fact, scientific researchers strive to develop theories because they are useful in predicting behavior. The importance of theories lies in their ability to predict an outcome or an event successfully. If a theory generates predictions that are supported by data, and the results are always the same, the theory can be used to make predictions in other similar situations.

MEDIA ANALYSIS STEPS

Evaluation of a problem must follow a standard sequence of steps to increase the chances of producing relevant data. Analysts who do not follow a prescribed set of steps increase the amount of error possible in the study.

These steps are:

- Select a topic (issue, situation, perception, or belief).
- Review existing research and other available information on the topic.
- Develop hypotheses and research questions.
- Determine an appropriate methodology, format or design.
- Collect relevant data.
- Analyze and interpret the results.
- Present the results in appropriate form (Information Paper, PA Study or PA Estimate)
- Validate and replicate the study when necessary.

The use of the scientific method of research is intended to provide an objective, unbiased evaluation of data pertaining to an issue or event. To investigate hypotheses systematically, media analysts must follow these steps. However, merely following the eight steps does guarantee that the research is good, valid, reliable or useful.

A countless number of intervening variables (influences) can destroy even the most well-planned research effort. Unanticipated events occurring during the research period may impact the results and they must be accounted for during the process. However, PA analysts must remain focused on the purpose of the research effort and not lose sight of the original objectives.

STEP 1 – SELECTING A TOPIC, DETERMINING RELEVANCE

Selecting a research topic is usually not a concern for Public Affairs analysts -- planning guidance, current situations, and most importantly, the operational issues confronting our commands, will guide the application of media content analysis. In most instances, the Public Affairs analyst will receive planning guidance well in advance of an operation, which will help determine the issues to be addressed.

Once the basic subject has been chosen, the next step is to ensure that it is relevant to the operation or situation at hand. This can be accomplished by answering six basic questions.

- What is the goal of this research effort?
- Is the subject too broad?
- Can the subject really be studied?
- Is the subject significant?
- Can the results of the research be generalized, communicated and understood?
- Does the issue lend itself to analysis?

Underlying all eight steps of the Media Analysis process is the necessity for validity. In other words, are all eight steps (from topic selection to data analysis to presentation and interpretation) the correct ones to follow in trying to answer these questions?

The answers to these questions will help focus the research you must do, make information gathering easier, and ensure the results are valid.

STEP 2 – REVIEW OTHER RESEARCH AND INFORMATION

Media analysts should always begin studies by consulting all literature, research, and other information available on the topic. The review provides information about what work has been done, how it was done, and what the results were as they apply to a given subject. It not only allows analysts to learn from (and eventually add to) previous media research, but also saves time, effort and money.

The review also helps to identify the facts pertaining to the situation being studied.

Completed media content analysis also provides a starting point for PA leaders who will follow in your position after you move on.

STEP 3 -- DEVELOP HYPOTHESES AND RESEARCH QUESTIONS

After the general research are has been identified and the existing information reviewed, the analysts must state the problem or issue as a workable **hypothesis** or **research question.**

(Example: "The American public is losing confidence in the Army's ability to protect its soldiers, resulting from the media's portrayal of Army leaders as negligent and soldiers as lacking competence in avoiding fratricide.")

A hypothesis is a formal statement regarding the relationship between variables and is tested directly. In the example cited above, those variables are the news media, the messages they send, and the perception and understanding of those messages by the American public. With a hypothesis, the predicted relationship between the variables is either true or false. Identifying the degree of "trueness" or "falseness" and their implications is essential to the development of information campaign strategies.

On the other hand, a research question is a formally stated question intended to provide indications about something, and is not limited to the relationships between variables. Research questions are generally used in situations where an analyst is unsure about the nature of the problem under investigation. The intent is merely to gather preliminary information. Research questions are generally used to identify the focus and scope of a research project.

STEP 4 -- DETERMINE AN APPROPRIATE METHODOLOGY/RESEARCH PROGRAM DESIGN

Given the variety of situations facing Public Affairs personnel, different approaches to media research are required. Some issues lend themselves more toward survey methodology via telephone, E-mail, or standard mail; others are best attacked through in-person interviews. Still other problems necessitate a controlled evaluation situation designed to eliminate extraneous variables by targeting analysis to specific media types. (An example of this approach would be a study of how a newspaper covered a specific story over a six-month period.)

The approach selected by the analyst depends on the goals and purpose of each particular study. Regardless of whether the problems or issue being addressed is a local one, affecting only a fraction of a community audience, or a national issue affecting us all, all research requires a design of some type. All procedures, including all variables, samples, and measurement instruments, must be pre-designed with hypothesis and research questions in mind.

There are four characteristics of research design that should be noted if a research study is to produce reliable and valid results:

- **Accurate setting**. For a study to have external validity, the study must be conducted as an historical account of the situation during the time frame studied. The analyst must have a clear understanding of the events unfolding around him and attempt to document as much related information as possible.
- **Clear cause-and-effect relationships**. The analyst must make every attempt to identify spurious dependent relationships and

weed them out. The results of a study can be interpreted with confidence if and only if all confounding effects are identified.

STEP 5 -- COLLECT RELEVANT DATA

STEP 6 -- ANALYZE AND INTERPRET THE RESULTS

STEP 7 -- PRESENT THE RESULTS IN APPROPRIATE FORM (PA STUDY OR PA ESTIMATE)

STEP 8 -- VALIDATE AND REPLICATE THE STUDY WHEN NECESSARY

Appendix P

INFORMATION PROGRAM EVALUATION

METHODS

How our audiences perceive the Army is critical to the success of all operations we are involved in. Internally, the Army's people require certain information to function effectively. The more they know and understand, the better they perform. Information about the operation, the unit's particular mission, how the commander feels about the situation, and a host of other subjects are of interest to both soldiers and civilians. Externally, the general public has specific needs for information about what their Army is doing and how they are doing it. This appendix explains methods for measuring success in the conduct of PA information programs.

Command information is communication between the commanders and those commanded. Command Information is different from the Public Affairs function of Information Provision in that it is the commander's responsibility to inform his people. Commanders must communicate their intentions and the troops, community, indeed the general public, must know his concerns and intentions. It is especially important to note that PA Information Provision techniques and procedures are just one channel that the commander may use in communicating to his audiences.

A poorly recognized fact is that the communication links between the commander and his audiences occur on various levels and assorted channels. This type of communication no longer fits the "top-down" communications model of the cold war Army. The explosion of today's digital technology has provided individual soldiers, civilians, family members, and the general public the ability to bypass rigid, controlled, vertical communication systems in favor of the common user, multidirectional, reciprocal, simultaneous, real-time transactive communications systems. Americans have the power to bypass the gatekeepers and ignore canned, shoddily produced, dated industrial age information products, in favor of accessing on-line information services or the Internet directly. Information is passed in all directions, continually. These audiences will have access to many more sources of information, which makes evaluating the effects of Army PA Information Programs all the more difficult.

This explosion of information technology has also highlighted how critical it is for PA elements to stay up to date on communications technologies, information services, and socio-economic trends of these forces at work.

Despite the rapid change in the information environment, the general steps for evaluating information program effects has remained the same:

- Determine the command's mission and the commander's method for accomplishing that mission.
- Identify all the various audiences interested in information related to the command, its members, and the mission.

- Identify a public opinion baseline -- the template against which new public opinion information will be compared.
- Gather identify all the messages communicated and identify which audiences' received such information.
- Gather information on the information program impacts. This is done reviewing unit newspapers, letters to editors, responses to information programs fact sheets, formations, surveys, and interviews. Check related bulletin boards on all on-line information services. Check related Newsgroups on world-wide-web nodes, which commonly carry related information. Monitor discussion groups on on-line services.
- Attend commander's calls, staff meetings, formations, briefings, and other gatherings where audience reaction, troop morale and like information will be discussed.
- Evaluate the knowledge of the targeted organizations. This is accomplished through in-person question/answer surveys and interviews, E-mail surveys, and electronic town hall meetings, etc.
- Coordinate with other staff elements addressing similar information issues (SJA, Chaplain, PMO, IG, etc.).
- Produce a summary of information gathered in an impact assessment.

Appendix Q

PRINCIPLES OF PA SERVICE AND STANDARDS

INTRODUCTION

1. This checklist identifies the primary tasks associated with the functions of Public Affairs, and establishes standards for successful accomplishment of those tasks. Standards of service equate to minimum exceptions of an operational commander and will be used to judge unit readiness, leader effectiveness, and individual soldier performance.

2. The following definitions apply to this checklist:

Austere: No existing PA units, assets or Army signal information infrastructure available in area of operations upon deployment. Commercial communications infrastructure is not available. PA elements must perform all missions and provide all PA support using organic personnel and equipment.

Existing: PA units, assets and Army signal information infrastructure in place before deployment. Commercial communications infrastructure is available. Deploying PA element assumes duties of or augments organic PA elements. Existing personnel and equipment augmented by additional PA elements to accomplish PA mission and provide PA support.

FUNCTIONS/TASK	CONDITION	STANDARD
Planning		
1. Perform Global Information Environment analysis	Existing	Austere
2. Develop PA Estimate (COAs)	Existing	Austere
3. Develop PA Strategy (plan)	Existing	Austere
4. Develop PA Guidance	Existing	Austere
5. Coordinate PA Annex	Existing	Austere
Media Facilitation		
1. ID media support requirements	Existing	Austere
2. Register news media	Existing	Austere
3. Coordinate media support	Existing	Austere
4. Provide media orientation	Existing	Austere
5. Coordinate news media interviews	Existing	Austere
6. Plan/coordinate news briefings	Existing	Austere
7. Establish Media OPS Center	Existing	Austere
8. Operate MOC	Existing	Austere

9. Provide media assistance/escort	Existing	Austere
Information strategy		
1. Identify target audiences	Existing	Austere
2. Identify information needs	Existing	Austere
3. Develop info themes/CMD messages	Existing	Austere
4. Gather info/develop products	Existing	Austere
5. Disseminate information to deployed forces	Existing	Austere
6. Disseminate information to families/home station audiences	Existing	Austere
7. Disseminate information to national/local news media	Existing	Austere
8. Disseminate information to general public	Existing	Austere
Public Affairs Training		
1. Identify training needs	Existing	Austere
2. Develop unit PA Training plan	Existing	Austere
3. Develop Family Spt Training Plan	Existing	Austere
4. Develop Senior LDR Training Plan	Existing	Austere
5. Conduct unit PA training	Existing	Austere
6. Conduct Family SPT Training	Existing	Austere
7. Conduct Senior LDR Training	Existing	Austere
8. Prepare SMEs for Media interviews	Existing	Astere
9. Evaluate training efforts	Existing	Austere

Appendix R

PRIVACY ACT/FREEDOM OF INFORMATION ACT

INTRODUCTION

This checklist addresses personal information about soldiers and any living persons that can or cannot be released under the provisions of the privacy act.

CHECKLIST ITEMS

- AGE (date of birth) = releasable. This information is public record.
- HOME OF RECORD/PRESENT ADDRESS = In most cases, home of record can be released if no street address is given. There is no general rule for disclosure of this information. Widely different circumstances surround each incident, and judgment is made on a case-by-case basis. In most cases, the person's present geographical location may be provided (city, state), but not the street address. In each case, the desires of the actual person or next of kin should be considered.
- MARITAL STATUS/DEPENDENTS = RELEASABLE. This information is public record, including names, ages and sex of dependents.
- AWARDS/DECORATIONS/CITATIONS = releasable.
- EDUCATION/SCHOOLING/SPECIALTY = releasable. Major area of study, school, year of graduation, degree and specialty designator is releasable.
- RACE = In most cases, NOT releasable. However, where the fact of an individual's race is relevant in providing essential facts to the Press, it may be released (such as in a racially oriented protest or altercation.)
- CHARACTER OF DISCHARGE:
- ADMINISTRATIVE = NOT releasable, unless the individual provides his written consent.
- PUNITIVE = releasable. This includes discharges resulting from courts martial.
- DUTY STATUS = releasable.
- PERSONNEL BOARDS:
- Results of promotion boards and augmentation boards are releasable.
- Results of administrative discharge boards and aviator flight boards are NOT releasable.
- PHOTOGRAPHS IN THE CUSTODY OF THE DEPARTMENT OF DEFENSE = releasable, unless they warrant an invasion of anyone's personal privacy.

Appendix S

PA GUIDANCE ON TERRORISM COUNTERACTION

In view of the continuing media interest in the subject of terrorism, the public affairs guidance contained in this message is provided to assist PAOs in responding to media queries and in developing local contingency plans.

TERRORISM IN GENERAL

a. **U.S. POLICY**. All terrorist acts are criminal. The U.S. Government will make no concessions to terrorists. Ransom will not be paid and nations fostering terrorism will be identified and isolated.

b. **RESPONSIBILITY**. Department of State is the lead agency for response to international terrorist incidents that involve U.S. military personnel and facilities outside the U.S. The administrator of the Federal Aviation Administration is responsible for terrorist incidents that affect the safety of DoD personnel or property aboard an aircraft in flight. When terrorist incidents occur at military installations within CONUS or its possessions (Puerto Rico, Guam, Virgin Islands, American Samoa, and trust territories), the FBI will be the lead agency. If the FBI declines to exercise its authority, military authorities may take appropriate action within the limits of their responsibility to resolve the incident.

c. **NOTIFICATION.** All terrorist incidents should be reported immediately through channels, to OASD/PA. No public release of information regarding a terrorist incident should be made without OASD/PA approval except for cases involving public safety.

COMBATING TERRORISM

Combating terrorism can be divided into two major areas: counter-terrorism (offensive measures) and anti-terrorism (defense measures).

- Counter-terrorism. The following statement may be used in response to queries regarding counter-terrorist forces:
- "The U.S. Government has trained forces and equipment from all four services to cope with terrorist incidents. We have also said that command and control elements for these forces exist and have been exercised. These elements report to the Joint Chiefs of Staff, as do other command and control elements for military operations. We do not comment on any details concerning the circumstances under which these forces may be deployed, their identity, or tactics."
- Requests for interviews or filming of counter-terrorism training will not, repeat will not, be approved.
- Requests for photos of counter-terrorist forces personnel or their training will not be approved.
- Because of the clear association/confusion surrounding the relationship between special operations counter-terrorism forces,

all requests for interviewing or filming special operations forces and or training will be approved by OASD/PA.

- Questions beyond the scope of the above guidance should be referred to OASD/PA..
- Anti-terrorism. The following guidance is applicable to media request for information pertaining to anti-terrorism.
- DoD officials, senior leaders, commands and knowledgeable individuals may discuss the subject of anti-terrorism as it pertains to those areas/installations for which they are responsible.(Anti-terrorism measures and procedures should be discussed in a general manner without going into a checklist of specific details.)
- Media requests to film anti-terrorist training will be approved on by OASD/PA.
- Photos of anti-terrorist training should be forwarded through channels for approval.

QUERIES

In response to queries regarding a possible or real terrorist threat at a particular base/installation/activity the PAO may acknowledge, if appropriate, that increased security measures have been/will be taken without going into specific details of all the measures taken. PAOs may, when appropriate, acknowledge the obvious.

For example, increased security measures such as increased guards at the gate or additional patrols, if they are obvious to the public may, in some cases, be acknowledged. PAOs should, however, exercise care and prudent judgment in any discussion of these or other security measures, which have been/will be implemented.

Appendix T

BROADCAST OPERATIONS

This section focuses on the radio and television services provided by Armed Forces Radio and Television Service at the unified command level and the coordination necessary by Army component commanders to ensure operational area support. It identifies and explains the AFRTS mission and its capabilities. It also discusses planning considerations and theater broadcast information requirements.

AFRTS CONTINGENCY BROADCAST OPERATIONS

The scope of the ABS mission of providing AFRTS radio and television news, information and entertainment programming to DoD personnel stationed overseas greatly expands during wartime to include support of global contingency requirements. As new contingency plans are developed based on emerging joint and Army doctrine, ABS must consider how the additional broadcasting personnel and equipment resources needed to support a rapid deployment broadcasting mission can be obtained while simultaneously meeting increased requirements in existing unified command theaters. The immediate response necessary to meet contingency requires the development of AFRTS appendices to Unified Command Operations Plans (OPLAN).

Army Broadcasting Service (ABS) is the Unified Command AFRTS Planner (UCAP) for the U.S. Southern Command and the U.S. European Command. ABS also has Geographic Area Planner (GAP) responsibilities for U.S. Forces Korea (USFK). ABS and AFRTS networks within these unified commands are responsible for updating and maintaining appendix content under provisions of the American Forces Information Service (AFIS) Concept Plan for Peacetime and Wartime Operations for the Armed Forces Radio and Television Service (AFRTS CONPLAN 98-1) and DoD Directive 5120.20-R

Contingency and/or Wartime Plans define the Mobilization/Contingency mission. Force structure to support these plans may be packaged as blocks, deployable units or detachments for ease of planning.

According to CONPLAN 98-1 and DoD Directive 5120.20-R, Unified Commanders and Subordinate Commanders, with the advice of ABS, determine the type of AFRTS Flexible Response Option (FRO) necessary. Unified and Subordinate Command support is required for the current levels of support contract and for any mission changes which affect AFRTS service in the CINC's area of operation.

AFRTS Mission

Provide live news, information and entertainment programming, free of censorship, to all DoD elements in place or deployed worldwide.

Provide U.S. military theater commanders with sufficient electronic media resources to effectively communicate DoD, Army, theater and AO command information.

Concept of Operations

Existing facilities and services will be the first AFRTS assets used to respond to AFRTS taskings.

Upon implementation of a Contingency Plan or OPLAN, AFRTS assets in the AO come under the direct operational control of the Unified Command for the period of the operation. When the operation is terminated, normal command relationships will be resumed.

The AFRTS Commander reports directly to the Unified Commander responsible for the theater of operations.

The AFRTS Commander retains direct command authority over AFRTS personnel and equipment.

The AFRTS Commander is responsible for all AFRTS matters concerning the operation, accomplishing direct coordination for the Unified Command with AFIS/AFRTS for all AFRTS issues requiring AFIS/AFRTS action, and managing all AFRTS assets involved in the operation in accordance with the AFRTS CONPLAN 98-1.

The AFRTS Commander will be a designated member of all public affairs meetings as a member of the staff.

The AFRTS Commander is authorized direct coordination with other members of the unified command's staff to work specific AFRTS support requirements.

All requests for command internal information or emergency announcements from AO organizations or personnel will be forwarded to the director of public affairs for approval.

All public information released by public affairs officials will be available for AFRTS use.

AFRTS radio and television electronic news gathering (ENG) will be dedicated for the AFRTS "on-air" mission and in direct support of the Unified Commander's internal information program.

AFRTS will provide full, factual and timely internal information and news to military audiences in the AO, consistent with national and operational security, and host country sensitivities.

AFRTS will follow Operations security (OPSEC) and communications security (COMSEC) rules.

AFRTS OPERATING PARAMETERS

AFIS is responsible for all AFRTS satellite programming services and overall policy and guidance for their use.

The AFRTS Broadcast Center (AFRTS-BC) is responsible for providing non-local and non-theater radio and television programming material to AFRTS facilities in the AO, except as outlined in Flexible Response Options (FROs).

The Television-Audio Support Activity (T-ASA) in Sacramento, California, is responsible for providing technical and logistical support to AFRTS.

The AFRTS facility in the AO will provide service based upon initial Flexible Response Options (FROs) and continue operations until directed to modify its services by AFRTS, the Unified Command or as the mission requirement dictates.

AFRTS will provide AO-wide announcements as required on both radio and television in order to facilitate unified command needs.

At unmanned repeaters and cable distribution systems, local officials may have AFRTS personnel make local announcements, if possible, in coordination with the AFRTS commander or on-site command representative, if approved by the director of public affairs for the operation.

AFRTS NETWORK COMMANDER OR DESIGNATED REPRESENTATIVE IN THE AREA OF OPERATIONS

The AFRTS network commander or designated representative will be co-located with the unified command director of public affairs.

The AFRTS network commander or designated representative will have a command function with direct operational command authority over all resources assigned to support the AFRTS mission in the AO.

The AFRTS commander or designated representative will ensure a logistics and engineering function responsible for providing advice and assistance to maintenance personnel assigned to AFRTS outlets, and maintaining unmanned equipment. This function will assist the AFRTS chief engineer in developing new equipment support requirements as changes occur in the AO.

The AFRTS network commander or designated representative will ensure internal information ENG coverage of unified command activities of interest to the members assigned in the AO. In joint service situations, Army AFRTS representatives may also be responsible for the production and duplication of radio and television internal information products for use at AFRTS outlets and television programming for DoD, or satellite cabled sites in the AO.

UNIFIED COMMAND RESPONSIBILITIES

The unified command provides logistic support for AFRTS. This includes vehicles, POL, and supply requirements in the AO as noted in the OPLAN. Also included is vehicle maintenance and POL for all AFRTS contingency vehicles. If additional forces are deployed to support AFRTS, the unified command assigns additional vehicles to support the expanded maintenance and production requirements.

If security and/or intelligence forces determine that AFRTS facilities have been identified as a potential target by hostile forces, the unified command will notify the AFRTS facility and provide security to targeted facilities.

If contract communications support is terminated during the implementation of the OPLAN, the unified command provides communications support for AFRTS use in the AO.

This includes existing long-wire or microwave systems for distributing the AFRTS signal and support for telephones, facsimile transmission and computer equipment for AFRTS.

If support cannot be obtained from existing assets, the unified command should be prepared to augment the AFRTS mission as outlined in the OPLAN.

The unified command provides personnel, administrative, vehicle and other logistic support for all AFRTS personnel assigned in the AO and those deployed to supplement that force. This includes unit line numbers and entry clearances required for all deployed personnel supporting the AFRTS mission.

The unified command obtains country clearance for construction of any temporary transmitter towers required due to expanded AFRTS service, which may occur as the operation unfolds.

The unified command provides electrical backup power for AFRTS facilities if contract services are terminated during the implementation of the OPLAN.

The unified command is responsible for obtaining necessary broadcast frequencies in consultation with the host-nation government to meet AFRTS broadcast requirements.

Flexible Response Options

Although each operation will differ, the following are general concepts of AFRTS Flexible Response Options (FROs) available for peacetime engagements, wartime operations and stability and support operations in an area where little or no AFRTS service exists or where crisis situations require a modification to existing AFRTS services. The unified command AFRTS planner (UCAP) is responsible for developing specific equipment, support and manning requirements to implement the AFRTS FROs that best support the specific operation.

The Unified Commander for the area of operations must request AFRTS radio and/or television services or for a change in present level of service before deployment. The Unified Commander requests AFRTS assistance through the unified command AFRTS planner responsible for the area of operations. The request will then be forwarded to ABS and AFIS for final approval.

FRO One: Direct to Ship (DTS) Service Support System. DTS is an U.S. Navy peacetime capability that provides news, sports, information and entertainment to audiences on ships at sea. A wartime adaptation of the service can provide immediate access to three radio and two television channels for land-based audiences including geographically separated units down to the lowest level. A deployable AFRTS kit containing an individual receiver decoder (IRD) provides service. This service provides a single-source 24-hour capability of receiving all services, but only one channel can be accessed at a time and no local or theater command information would be available.

FRO Two: Satellite Direct Radio and Television (SDRTV). SDRTV is an unmanned AFRTS satellite service that can be provided to virtually any land based audience on the globe with up to 10 stereo radio channels and six television news, sports, information and entertainment channels. In the European theater it will include a regionally generated signal. Service is provided at a single location using a deployable AFRTS kit containing a simultaneous receiver decoder (SRD) that provides a capability of receiving all services and accessing up to six radio and television programming sources at a time. As a public affairs option,

SDRTV provides an internal information data stream that can be accessed with the addition of a computer, printer and proprietary software to the SDRTV equipment package for use by public affairs activities in providing support to deployed populations. Unified command public affairs offices and the UCAP should consider coordinating the use of the additional capability whenever a manned public affairs activity is deployed.

FRO Three: Manned Radio Systems. Signal Distribution Systems. These deployable systems include audio and video transmission and cable systems that provide a capability to distribute, DTS, SDRTV or manned radio service to an expanded autonomous geographic area such as a base camp or Air Force base in an AO.

FRO Four: Manned Radio Systems. These deployable systems provide a capability for local, live internal information and radio news. Various types and sizes of local radio systems exist that can be used to establish a range of services from simple local break away "radio-in-a-box" to a full service facility with local production capability. Some of these systems will include radio transmitter that can provide limited signal distribution without deployment of FRO Three. This system can provide a limited single-source radio service to outlying populations that are not served by FRO One or Two, are not available or would not be appropriate programming sources.

FRO Five: MOOTW Management, Local TV and Network Live Radio. These deployable personnel and systems support the development of an AFRTS management function to oversee dispersed AFRTS operations and will add local television and network-wide live radio capability using organic distribution systems. The capability will establish a network to support operations in an AO comprising a large peacetime engagement of medium to long duration encompassing a large geographic area where the operations commander requires near real-time internal information capability. The system provides network administrative, computer, maintenance, engineering and operations support functions. These functions provide the unified command director of public affairs with AFRTS management expertise not normally available on the public affairs staff. The television service system is designed to produce AO information that can also be fed to the AFRTS Broadcast Center for rebroadcast to DTS/STRTV audiences worldwide.

FRO Six: Theater Satellite Radio and Television Operations (TSRTO). In a major regional conflict where large force deployments are planned the AFRTS Broadcast Center will dedicate one channel of radio and one television channel for use by the UCAP to broadcast directly to the theater of operations. Programming will include time shifting "prime time" so that each 12-hour shift receives prime time programming in the first four and-one-half hours of off time. In conjunction with FRO One or Two, this will provide a virtual network capability to the AO commander and the PAO. It will appear to the audience as if the broadcast was occurring in the AO when in fact it is originating from the Broadcast Center. Initial spots can be unsophisticated radio readers and character-generated (CG) messages on television. If there is a manned radio facility providing theater and operation-specific internal information in place, they will forward copies of all AO specific spots via computer to the Broadcast Center for use initially as television CG messages within the dedicated TV channel. As with all internal information, they will be developed in coordination with the PAO in the AO. If there is insufficient AFRTS manning in place which would be the case if there were numerous geographically-separated operating locations, the internal information will be supplied, via computer, by the AO public affairs office. Base and component command PAOs will be advised of the scheduling of theater programming and encouraged to provide service-unique spots for use in theater. Service will continue until a significant drawdown occurs, the operation stabilizes to the point where local TV spots are not needed and the presumption of pre-conflict programming will serve the majority of the deployed population or an AO based network begins serving the operation.

Appendix U

COMMUNITY SURVEY

COMMUNITY SURVEY EXAMPLE

I.The area

A. Geographical description
1. Of areas surveyed--size of cities, counties.
2. Of surrounding area, if pertinent.
3. Climate, topography, annual and seasonal temperatures, rainfall, etc. (one sentence will suffice for each.)
4. Are the industries dispersed or centralized? Attach a map of the Area indicating the location of the principal plants. The map should show the names and numbers of principal streets and highways furnishing access to these plants.

B. Population
1. Of city
2. Of area.
3. Of labor market area, if different from above.
4. Breakdown by sexes, color, native or foreign born, educational level, percentage of homeowners, etc.

C. Industrial data
1. Types of industries and number of each, labor force of each, key products of the area, and additional data as considered applicable.
2. Does one type of industry dominate the area? If so, give pertinent information regarding the industry.

II.Manpower

A. Labor market rating
1. Is department of labor market classification, a,b,c, or d?
2. Include supporting statistical data.

B. Unemployment
1. Totals and percentages of skilled, semiskilled, and unskilled.
2. What types of skills are most commonly available?

C. Source of labor supply

1. Compared to the World War II years, what are the reserves of women, handicapped, older-age groups, part-time workers, and school graduates?

2. Has there been much intermigration to total population?

D. Occupational classification of area workers. What are the most common occupations of the area? The less common?

E. Skills in shortage category. List, with numbers of each, if available. Make a comparison of this list with the national shortage list.

F. Area wage schedules
   1. List the wage schedules of major occupations and industries.
   2. How do they compare with national averages?
   3. How do they compare with neighboring areas? With competing areas?

G. Requirement of defense industry in area
   1. Is manpower available for present production schedules? Current planned production?
   2. What skills are lacking for production schedules, both present and future?
   3. Do employers ordinarily use training programs? If so, give some Examples.

H. Other pertinent information
   1. Include current work stoppages, if any; record of work stoppages During last 10 years.
   2. Are workers highly organized? Principal unions?

III.Industrial facilities

A. Facilities suited or adaptable for defense production.
B. List facilities with current and World War II products.
C. Give current and capacity employment.
D. Give types of machinery.
E. What defense contracts are held or sought?
F. Vacant factory space. Describe space and indicate production potential.

IV.Housing

A. Housing regulations.
   1. Is it a critical defense housing area under public law 96? P. L. 139?
   2. Local rent control?

B. Housing units available.
   1. Number for sale, including 1, 2, and 3 bedrooms. Price ranges. Are the prices reasonable?
   2. Number for rent, including above information. Apartments available. Number, size, price ranges. Are the rents reasonable?

3. Sleeping rooms available. Number, price ranges. Are the prices reasonable?
4. Building permits issued (in past 12 months)
5. Number for houses--1, 2, 3, or more bedrooms.
6. Number for apartments.
7. If houses, number for rent and contemplated rental prices.
8. Housing units contemplated .
9. Number and sizes. How many bedrooms?
10. Number of these for rent.
11. Estimated rental rates.

C. Builders
1. Adequate number of experienced builders?
2. Do they have trained skeleton force?
3. Can other necessary housing construction workers be secured?
4. Is land available? Under option?
5. Can materials be obtained?
6. What bottlenecks?

D. Building capital.
1. Is capital for the building of housing and rental units readily available? If so, on long- or short-term loans?
2. What are the sources of this capital?
3. Does the community object to construction of more housing units now?

V.Adequacy of housing

For present work force? For expanded production? (quote a figure or percentage, such As peak load in world war ii or 50 percent above present.) This figure should be adequate to cover planned defense expansion known to you at time of Survey.

VI.Other community facilities and services

Discuss each of the following items as to adequacy for the present work force and for an expanded work force; give specific facts for each, as pertinent (yes and no answers are not adequate.)

A. Water
B. Electric power
C. Gas
D. Sewerage

*Note: for items 1 through 4 above, describe sources of supply, capacity, reserve storage, current use, reserves on hand, plans for expansion--whether on hand or projected.

E. Transportation: types and numbers
F. Highway and road systems: are the roads serving the area adequate and in good? What is the present traffic load and the peak capacity of these roads? Describe any unsatisfactory factors. What action is contemplated or considered necessary to assure free traffic movement within the area? (survey requests will furnish, whenever possible, specified information on any industrial and

defense manpower requirement changes under consideration for the area.) Contact with local, state, and federal highway authorities should be made, if necessary, to explore fully this phase of the survey.

1. Schools: number of each type of school, crowding, shifts, new construction, etc.
2. Hospitals: number, number of beds, population per bed.
3. Doctors: number, population per doctor and per dentist.
4. Fire protection: size, ratios, and ratings.
5. Police protection: size, ratios, and ratings.
6. Shopping centers and shopping hours.
7. Recreational facilities: number of each type.
8. Churches: all denominations.
9. Sanitation service (garbage collection).
10. Laundries, dry-cleaning businesses, barber shops, beauty shops, etc.
11. Banking facilities (include arrangements for shift workers).
12. Hotels: number and number of rooms, scale of rates, etc.
13. Restaurants and other eating places.
14. Newspapers: number (morning, evening).
15. Municipal government (form, etc.).
16. Tax rate: local, county, state.
17. Cost of living index: get whatever information is available Compare local figures to national index.

*Note: include reference material, maps, booklets, etc., If possible.

Appendix V

AUDIENCE SURVEY

Audience surveys systematically gather information about the effectiveness of CI programs and products as they relate to a particular group of people. The commander and the PAO to make decisions about management and direction of an internal information program or product use the results.

PUBLIC AFFAIRS REQUIREMENT

The PAO will conduct a readership survey at least every two years (every three years for the Reserve Components). Coordination with the DOIM for possible computer and analysis support is recommended. Additionally, AR 600-46 can provide information on conducting surveys. The survey will provide data on distribution effectiveness, reader awareness and acceptance, readership and perceived usefulness of standing features and topics covered, and opinions of the value and effectiveness of the publication. Repeated surveys will provide trend data.

The survey may include any or all of the 20 questions listed in the Readership Survey (RCS: SAOSA-223) (app H) in AR 360-81. However, surveys not using these tested questions must be pretested to ensure validity before being used in a survey. Survey respondents will be selected using probability-sampling techniques.

Informal surveys, such as those included in a newspaper or conducted randomly/haphazardly with a few people, are not substitutes for readership surveys. This does not preclude an editor from periodically publishing a coupon or set of questions to solicit informal feedback that is not statistically projectable.

Before administering the survey, the survey managers must coordinate with the agency that will provide response analysis to be sure questionnaires; answer sheets, data entry program, or any other materials are appropriate and usable. Survey conduct may be included in the command's CE publication contract and may also be contracted by the command for Army Funded newspapers, providing funds are available.

When civilian employees are surveyed, PAOs should also coordinate with the civilian personnel officer for local union notification requirements. Completed questionnaires may be analyzed by the local Director of Information Management (DOIM) to provide percentages of responses to survey questions. Where computer support is available (from the local DOIM or DRM), responses will be analyzed using a program such as the Statistical Program for the Social Sciences package. Questionnaires must be constructed using the parameters of available software.

A written discussion of findings and conclusions drawn from the survey will be forwarded within 60 days after the survey is completed through the local commander, appropriate major command, to HQDA (SAPA-CI-PMN), Room 2E625, The Pentagon, WASH DC 20310-1510.

As a minimum, the report will contain the survey statistics, an analysis of the data, identification of strengths and problem areas (e.g., distribution, more sports, etc.), recommended improvements and changes to editorial policy, and an indication that the commander has reviewed the results.

Surveys may be conducted any time. However, no newspaper's survey report on file at HQDA should be older than 3 years (4 years for the Reserve Components). This allows for the time to conduct a survey.

A copy of the most recent survey will also be submitted with the annual CI Program Assessment Report (DA Form 510-R), unless the survey was previously submitted to the MACOM and OCPA-HQDA.

The PAO will conduct electronic media surveys at least every two years (every three years for the Reserve Components).

SURVEY ADMINISTRATION

Among the more common methods of conducting surveys are the mail survey, face-to-face interview, and telephone interview. The mail survey is the preferred method for purposes of this requirement, although other methods, managed properly, may be used.

Survey respondents will be selected in a totally random manner (e.g., simple random, stratified, or systematic, using probability sampling procedures). Survey managers will select samples, which achieve at least a minimum of a +/- 5 percent reliability (error margin) at the 95 percent confidence level.

Sample sizes shown for the various reliability levels (e.g., +/- 5 percent error margin) are the number of usable responses received, not the number of questionnaires to be sent out. For a population of 5,000, 357 usable responses will accurately reflect, to within +/- 5 percent, what the entire 5,000 member audience would have said, had it completed the survey.

Experience with mail audience surveys shows that they realize an approximate 30 to 35 percent response rate. Therefore, send out at least three times as many questionnaires as are needed for analysis. Remember that incentives encourage responses.

Maybe the local MWR office or similar staff agency could provide bumper stickers, discount coupons, or other incentive for completed responses.

FOCUS GROUPS

Focus groups. One of the most effective ways to learn how CI products are being received is to conduct focus group interviews. Focus group interviews are structured group discussions in which representative members of the audience are brought together to discuss one or more CI products or issues. These interviews or sessions can examine the effectiveness of products or programs, gain suggestions for improving existing products or programs, and determine the need for new products or programs.

The key to effective focus group interviews is proper planning. Focus group organizers must determine who will participate, and what are the specific objectives of the session (e.g., what topics or issues will be discussed, what specific questions will be asked, what is to be done with the results, etc.). The method of selecting participants should be determined and the location for the meeting secured. While there is no optimal size for focus groups, generally groups of six to 10 individuals are manageable. Group makeup (officer/NCO/enlisted, men/women, military/civilian, active/reserve, retired/family member) depends on the objectives of the session. Generally, homogeneous groups are preferable. Often it will be necessary to hold more than one focus group session to obtain information needed to evaluate a particular CI program or product.

The moderator or group leader should be someone skilled in interview techniques and knowledge about the product or program being evaluated. It is often best not to have a high-ranking individual as the moderator with a group of junior enlisted or young family members, as free flow of information and opinions may be inhibited. The group leader must facilitate the discussion, not serve as an interrogator.

Focus group sessions should be informal. Participants should be encouraged to speak whenever they wish; the moderator should focus the discussion on the topics without being overbearing. If participants agree, it will be useful to videotape their comments for use in evaluating the session.

It is important that all group members understand that their honest opinions are being sought, and that the session is intended as a positive method of improving CI within the organization. No punitive actions should occur as a result of these sessions.

Appendix W

PA Lessons Learned

"What this century's history teaches us is that the Army's real strength is its ability to change and adapt to the period's requirements. Our ability to change was the key to victory in two world wars and a cold war, and it will be the foundation for our future success."

--General Dennis J. Reimer

INTRODUCTION

Explosive developments in information age technology have made the prospect of sharing lessons and ideas across a wide audience a reality today. With ready and easy access to E-mail and the Internet, soldiers can distribute documents, graphics, and photographs with lightning speed.

This appendix is based on an article published in the Center for Army Lessons Learned (CALL) ***News from the Front!***

This section provides public affairs officers (PAOs) with a tool for capturing observations and an outlet for rapid analysis and dissemination of tactics, techniques and procedures (TTPs) to the force. The initial focus is on defining and narrowing the scope of TTPs. It is important for officers and soldiers in the field to clearly understand the process before proceeding to methodologies of collection. The next section provides a structure for developing a narrative product for publication by CALL. Such a product will provide a coherent article of information, which can be quickly used by the force. The final section describes the observation-gathering process. Understanding the process for collecting data will prove invaluable to the operational planner and for producing effective training tools for the future.

DEFINITION AND LIMITS

Focusing the collection effort is central to capturing meaningful observations. Although CALL regularly sends combined arms assessment teams (CAATs) to major exercises and actual operations to gather observations, units from the field, including any public affairs section or detachment, can provide great insight by planning for the collection of information. In fact, only the Army as a whole can make CALL a continuing conduit of information for use by soldiers.

By using the structure and tools described below, units can provide useful TTP by establishing a collection effort as part of the originating operation order (OPORD), with almost no interference with normal operations. Indeed, the tools will enhance planning for future (and remedial) training by incorporating the capturing of TTP into the plan.

TTP are often limited to the specific operation or exercise. The function and use of TTPs are analogous to legal precedents. In law, if given circumstances of a case are generally similar to a prior case, it is assumed that a judgmental decision for the present case should be the same. However, circumstances in law often have aspects that are unique and must be considered before rendering a new decision. When applying TTP, study prior situations in context and use the lessons prudently.

STRUCTURING THE PAPER FOR USE BY CALL

PAOs at all levels can build upon the after-action review (AAR) process in the plan by producing a publishable document. In almost all exercises, units learn and consequently implement improvement measures. By employing the structure below, units can effectively share information throughout the force -- not only from Combat Training Center (CTC) rotations but also from home-station training and exercises away from the training centers.

Do not view the structure below as a rigid construct. Rather, it should serve as a point of departure for unit writers. Although quantitative material is useful for commanders and researchers, make this document narrative in format. Use graphics to support the narrative, if possible. Bring together data into a cohesive product that other units can readily use without resorting to sifting through large amounts of charts, lists, and disjointed bullets.

- Type of unit. Describe the type of unit the PAO supported (mechanized infantry division, separate brigade).
- Context of event. Summarize the general setting for the exercise or operation. (See Exercises and Actual Operations below. More operational context information is provided in this section.)
- Commander's comments. If possible, the commander can provide a brief (one or more paragraphs) commentary on public affairs operations. Work closely with the unit's executive officer or chief of staff for such input.
- Interaction with PSYOP, Civil Affairs, Signal. As information operations continues to grow and doctrine is further developed, interaction between various agencies will also continue to expand. While ensuring coordination with PSYOP and civil affairs operations, PAOs will continue to recognize the separation in functions of the organizations required by law. Discuss the coordination measures used.
- Media Relations.
- Summary of events. Provide a summary of events. Were press conferences and interviews scheduled and executed? What was the pace of daily operations? What was the routine daily schedule?
- Command messages. In developing this section, answer the following questions in detail: What were the command messages? More importantly, did the command messages come through to print or broadcast? Were any command messages distorted or misinterpreted? How can clarity be improved for the next operation?
- Summary of higher headquarters' public affairs guidance (PAG). Write a one or two paragraph summary of the initial and follow-up PAG received from higher headquarters. (Provide a complete copy as an appendix.) Provide answers to the following questions

following the summary: How did PAG influence operations? Was the PAG clear and meaningful? Were excerpts used to create lower level command messages?

- Media contacts. Describe the types and numbers of media contacts. Did the unit encounter numerous print-journalist requests? Electronic requests? Were there patterns in the requests? What could a future media preparation package contain to answer some questions in advance? How did the PAO prioritize media access? Were major outlets afforded more opportunities?
- Summary of Media Releases. Summarize media releases in one or two paragraphs. What were the major themes? What media received the releases? Were releases used in stories? Were there any comments from members of the media about the releases?
- Media Content Analysis. During and following an event, gather press clippings and, if possible, record electronic media stories about the event. What was the nature of the coverage? What was the tenor of editorial comments? Did command messages get exposure? Was the content of articles generally accurate? What could PAOs do in the future to improve the accuracy of content?
- Command Information Products. The command information program in the field is fundamental in the minds of American

CIVIL WAR

"When the (Civil) war entered Pennsylvania, the Philadelphia Inquirer often sold up to 25,000 copies of a single issue to the men in the field. During a lull in the Battle of Cedar Creek in October 1864, observers later remarked that the first thing the men did along the line was to sit down, boil coffee, and pull out their newspapers."

soldiers. And, this phenomenon is nothing new to this culture.

- Because of this intense internal interest in events surrounding the operation and events back home, it is imperative that PAOs adequately address methods and practices used to inform the soldiers. Answer the following questions in the narrative: Did soldiers in the field receive consistent and timely information about operations and world events? What was the distribution method? How was it evaluated for effectiveness? Were reproduction resources available and used adequately?
- Changes Incorporated in the Tactical Standing Operating Procedures (SOPs). From the TTPs gathered, what changes will the unit make to the SOP? Briefly describe why the unit is making the change and what mechanisms to put in place to test the effectiveness of the change.

EXERCISES AND ACTUAL OPERATIONS

Operational information is important for the reader to understand the context in which the PAO operated. Provide the following information for the CALL document to contribute to the reader's full understanding of the event(s):

- Mission - Summarize the unit's mission. (The focus here is not on the PAO's mission, but on the mission of combat headquarters.)
- Commander's Intent - This information is available on the OPORD. By incorporating this information, the reader will have an appreciation of the context in which the PAO operated. Emphasize the components of the intent:
    - Purpose
    - Method
    - Endstate
- New Equipment Used - Was new equipment available and used in the operation? Describe the equipment. Was it useful? What were the additional training requirements for using the equipment?
- New Techniques Used - Did the PAO incorporate techniques which are not described in doctrinal manuals? Describe the techniques employed.
- Structure - What was the structure of the PAO unit or shop? What manning -- required versus on-hand? (Note: Do not provide information that is classified under provisions of unit strength reporting (USR) regulations. Seek to provide a document that is free of classified material.)
- Operational Developments - How did the headquarters change its plan during the operation? How did the change(s) affect PAO operations? The descriptions here will bridge the gap between the original plan and its actual implementation.
- TTPs applied during the mission and for future operations - In bullet narrative, describe TTPs gathered in the operation. The bullets must contain sufficient detail for the reader to understand the situation and application possibilities for future operations. Support the bullets by providing individual observations (see TTP-Gathering Process below) as a combined appendix. The narrative in the base document must stand alone, with the appendix of individual observations providing additional detail.

TTP-GATHERING PROCESS

Units can contact CALL when developing plans for collecting TTP. CALL analysts can provide observer guidance, assist in delineating responsibilities of observers, identify documents or reference for use in developing a collection plan, and describe collection methodology.

- Observations. Individual observations assist in providing the basis for the narrative document described above. Use the form below to capture observations and develop a database for use in narrative development. Provide a copy of each observation to CALL as an appendix to the narrative. (Note: Any document published by CALL will not list units nor individuals by name. Refer to units by level ("the division" instead of "the 101st Airborne Division") and personnel by position ("a brigade chaplain" instead of "Chaplain Jones"). The purpose of CALL publications is to share ideas - not to point fingers.

- Observation Forms. The observation form (Appendix A) can be used for individual observations. A Microsoft Word version is available. Contact CALL via E-mail at call@leavenworth.army.mil or DSN 552-9571 (commercial 913-684-571) to receive a copy of the document. The document contains key components which aid the researcher in preparing analyses:
- Observer Name - The observer's name is used administratively only. No observer's name will appear in a CALL product.
- Administrative Information - Like the observer's name, unit information is used administratively only. Unit names do not appear in CALL products.
- Observation Indicators - Check all the appropriate blocks.
- Interoperability Indicators - Check all the appropriate blocks.
- Environmental Indicators - Check the appropriate block.
- File Name - Employ a system that differentiates each observation. One method is for observers to use name initials combined with sequence number and date (John Smith's first observation of May 5th would read, jsmay0105). Other systems are acceptable if plainly explained.
- Observation Title - Give the observation a brief, distinct title.
- Observation - In one sentence, summarize the observation.
- Discussion - Provide as much detail as necessary to provide a clear picture to the analyst or future reader. The length of the discussion will vary.
- Lesson Learned - In the context of your observation, provide a TTP.
- DTLOMS Implications - Describe how the observation impacts one or more areas in DTLOMS:
  - Doctrine
  - Training
  - Leadership Development
  - Organization
  - Materiel
  - Soldier Support

Include other media support, such as photos, sketches, or slide presentations in support of the narrative text.

Provide the narrative text, appendices and other material to CALL at the following locations: call@leavenworth.army.mil, or Department of the Army, Center for Army Lessons Learned, 10 Meade Avenue, Fort Leavenworth, Kansas 66027-1350.

In addition to maintaining and expanding a database of information, CALL publishes ***News From the Front!***, and a host of other publications for easy use by the force. Importantly, much of what is available has originated from the field -- from soldiers just like you. ***News From the Front!*** is published six times per year and provides a forum for a wide variety of topics of interest to the field. Other publications, including newsletters, CTC Bulletins, special editions, Handbooks, and more, focus on specific topics. Much of the published holdings of CALL can be found on the CALL website, http://call.army.mil, post libraries, or by contacting

CALL at the E-mail address listed above. Various search engines are available on the website to assist researchers.

CONCLUSION

Sharing information is possible with rapid and potentially colossal results. Leaders and soldiers who understand the TTP-gathering process can build plans for the future into every OPORD. By incorporating a plan to collect data and produce a clean narrative product for use by the force, soldiers throughout the Army gain maximum benefit from existing and future advancements in information technology. PAOs can focus on critical elements for successful media relations operations and command information programs in the field. Planners can easily adapt collection plans to exercises or actual operations -- anywhere in the world.

Learning is crucial for continued success on the battlefield. PAO planners must inculcate a practice of gaining a full understanding of the process and incorporating it into future exercises and actual operations.

Appendix X
MEDIA GROUND RULES

OPERATION DESERT STORM GROUND RULES

The following information should not be reported because its publication or broadcast could jeopardize operations and endanger lives:

(1) For U.S. or coalition units, specific numerical information on troop strength, aircraft, weapons systems, on-hand equipment or supplies (e.g. artillery, tanks, radars, missiles, trucks, water), including amounts of ammunition or fuel moved by support units or on hand in combat units. Unit size may be described in general terms such as "company-size, multi-battalion, multi-division, naval task force and carrier battle group." Number and amount of equipment and supplies may be described in general terms such as "large, small, or many."

(2) Any information that reveals details of future plans, operations or strikes, including postponed or cancelled operations.

(3) Information, photography and imagery that would reveal the specific location of military forces or show the level of security at military installations or encampments. Locations may be described as follows: all Navy embark stories can identify the ship upon which embarked as a dateline and will state that this report is coming "from the Persian Gulf, Red Sea or North Arabian Sea." Stories written in Saudi Arabia may be datelined "Eastern Saudi Arabia, near the Kuwaiti border, " etc. For specific countries outside Saudi Arabia, stories will state that the report is coming from the Persian Gulf region unless that country has acknowledged its participation.

(4) Rules of engagement details.

(5) Information on intelligence collection activities, including targets, methods and results.

(6) During an operation, specific information on friendly force troop movements, tactical deployments and dispositions that would jeopardize operational security and lives. This would include unit designations, names or operations and size of friendly forces involved until released by CENTCOM.

(7) Identification of mission aircraft points of origin, other than as land or carrier based.

(8) Information on the effectiveness or ineffectiveness of enemy camouflage, cover, deception, targeting, direct and indirect fire, intelligence collection or security measures.

(9) Specific identifying information on missing or downed aircraft or ships while search and rescue operations are planned or underway.

(10) Special operations forces methods, unique equipment or tactics.

(11) Specific operating methods and tactics, (e.g. air ops angles of attack or speeds, naval tactics and evasive maneuvers). General terms such as "low" or "fast" may be used.

(12) Information on operational or support vulnerabilities that could be used against U.S. forces, such as details of major battle damage or major personnel losses of specific U.S. or coalition units, until that information no longer provides tactical advantage to the enemy and is, therefore, released by CENTCOM.

*** Damage and casualties may be described as "light," "moderate," or "heavy."**

Glossary

AFRTS— Armed Forces Radio and Television Services

BOD—Broadcast Operations Detachment

C2 protect—command and control-protect—see **command and control warfare**

C2W—command and control warfare

command and control warfare—The integrated use of operations security (OPSEC), military deception, psychological operations (PSYOP), electronic warfare (EW), and physical destruction, mutually supported by intelligence, to deny information to, influence, degrade, or destroy adversary command and control capabilities, while protecting friendly command and control capabilities against such actions. Command and control warfare applies across the operational continuum and all levels of conflict. Also called C2W. C2W is both offensive and defensive:

C2-protection—To maintain effective command and control of own forces by turning to friendly advantage or negating adversary efforts to deny information to, influence, degrade, or destroy the friendly C2 system.

civil affairs—the activities of a commander that establish, maintain, influence, or exploit relations between military forces and civil authorities, both governmental and nongovernmental, and the civilian populace in a friendly, neutral, or hostile area of operations in order to facilitate military operations and consolidate operational objectives. Civil affairs activities (1) embrace the relationship between military forces and civil authorities and population in areas where military forces are present; and (2) involve application of civil affairs functional specialty skills, in areas normally the responsibility of civilian government, which enhance conduct of civil-military operations. These activities may occur prior to, during, or subsequent to other military actions. They may also occur, if directed, in the absence of other military operations.

combined operation—an operation conducted by forces of two or more allied nations acting together for the accomplishment of a single mission.

command information—see **internal information**

community relations—establishing and maintaining effective relationships between military and civilian communities through planning and active participation in events and processes which provide benefits to both communities.

community relations program—that command function which evaluates public attitudes, identifies the mission of a military organization with the public interest, and executes a program of action to earn public understanding and acceptance. Community relations programs are conducted at all levels of command, both in the United States and overseas, by military organizations having a community relations area of responsibility. Community relations programs include, but are not limited to, such activities as liaison and cooperation with associations and organizations and their local affiliates at all levels; armed forces participation in international, national, regional, state, and local public events; installation open houses and tours; embarkation in naval ships; orientation tours for distinguished civilians; people-to-people and humanitarian acts; cooperation with government officials and community leaders; and encouragement of armed forces personnel and their dependents to participate in activities of local schools, churches, fraternal, social, and civic organizations, sports, and recreation programs, and other aspects of community life to the extent feasible and appropriate, regardless of where they are located.

GIE—global information environment

global information environment—all individuals, organizations, or systems, most of which are outside the control of the military or National Command Authorities, that collect, process, and disseminate information to national and international audiences.

ground rules—conditions established by a military command to govern the conduct of news gathering and the release and/or use of specified information during an operation or during a specific period of time.

information age—the future time period when social, cultural, and economic patterns will reflect the decentralized, nonhierarchical flow of information.

information architecture—Description and specifications of information systems to include identification of communicators, information transmitted, equipment specifications and network designs. Includes operational, system and technical architectures.

information operations—continuous military operations within the military information environment that enable, enhance, and protect the friendly force's ability to collect, process, and act on information to achieve an advantage across the full range of military operations; information operations include interacting with the global information environment and exploiting or denying an adversary's information and decision capabilities.

information strategy—a synchronized plan for using all available and appropriate methods of communication to achieve specific goals of informing target audiences.

internal information—communication by a military organization with service members, civilian employees and family members of the organization that creates an awareness of the organization's goals, informs them of significant developments affecting them and the organization, increases their effectiveness as ambassadors of the organization, and satisfies their desire to be kept informed about what is going on in the organization and operation (also known as command information).

JIB—joint information bureau

joint information bureau—facilities established by the joint force commander to serve as the focal point for the interface between the military and the media during the conduct of joint operations. When operated in support of multinational operations, a joint information bureau is called a Combined Information Bureau or an Allied Press Information Center.

joint force—a general term applied to a force composed of significant elements, assigned or attached, of two or more Military Departments, operating under a single joint force commander.

joint operations—a general term to describe military actions conducted by joint forces, or by Service forces in relationships (e.g., support, coordinating authority), which, of themselves, do not create joint forces.

JOPES—Joint Operations Planning and Execution System

joint task force—a joint force that is constituted and so designated by the Secretary of Defense, a combatant commander, a subunified commander, or an existing joint task force commander. Also called JTF.

media facilitation—the range of activities such as providing access and interviews that assist news media representatives covering military operations.

media operations center—facility that serves as the focal point for the interface between the military and the media covering an event or operation.

media pool—a limited number of news media who represent a larger number of news media organizations for news gathering and sharing of material during a specified activity. Pooling is typically used when news media support resources cannot accommodate a large number of journalists. The DoD National Media Pool is available for coverage of the earliest stages of a contingency. Additionally, the combatant commanders may also find it necessary to form limited local pools to report on specific missions.

METT-TC—mission, enemy, terrain, troops, time available and civilians

MIE—military information environment

military information environment—the environment contained within the global information environment, consisting of the information systems and organizations—friendly and adversary, military and nonmilitary—that support, enable, or significantly influence a specific military operation.

MPAD--Mobile Public Affairs Detachment

multinational operations—a collective term to describe military actions conducted by forces of two or more nations, typically organized within the structure of a coalition or alliance.

news media representative—an individual employed by a civilian radio or television station, newspaper, newsmagazine, periodical, or news agency to gather information and report on a newsworthy event.

NGO—nongovernmental organization

nongovernmental organizations—transnational organizations of private citizens that maintain a consultative status with the Economic and Social Council of the United Nations. Nongovernmental organizations may be professional associations, foundations, multinational businesses, or simply groups with a common interest in humanitarian assistance activities (development and relief). "Nongovernmental organizations" is a term normally used by non-United States organizations. Also called NGO. See also private voluntary organizations.

PAD—Public Affairs Detachment. The smallest of the PA units.

PAG—public affairs guidance

PAOC—Public Affairs Operations Center.

private voluntary organizations—private, nonprofit humanitarian assistance organizations involved in development and relief activities. Private voluntary organizations are normally United States-based. "Private voluntary organization" is often used synonymously with the term "nongovernmental organization." Also called PVO. See also nongovernmental organizations.

psychological operations—operations to convey selected information and indicators to foreign audiences to influence their emotions, motives, objective reasoning, and ultimately the behavior of foreign governments, organizations, groups, and individuals. The purpose of psychological operations is to induce or reinforce foreign attitudes and behavior favorable to the originator's objectives. Also called PSYOP.

PSYOP—psychological operations

public affairs assessment—an analysis of the news media and public environments to evaluate the degree of understanding about strategic and operations objectives and military activities and to identify levels of public support. Includes judgments about the public affairs impact of pending decisions and recommendations about the structure of public affairs support for the assigned mission.

public affairs estimate—as assessment of a specific mission from a public affairs perspective.

public affairs guidance—normally, a package of information to support the public discussion of defense issues and operations. Such guidance can range from a telephonic response to a specific question to a more comprehensive package. Included could be an approved public affairs policy, news statements, answers to anticipated media questions, and community relations guidance. Public affairs guidance also addresses the method(s), timing, location and other details governing the release of information to the public.

public information—A general term describing processes used to provide information to external audiences through public media.

sustaining base—the home station or permanent location of active duty units and Reserve Component units (e.g., location of armory or reserve center) that provides personnel, logistic and other support required to maintain and prolong operations or combat.

Bibliography

Joint Pub 3-13. Joint Doctrine for Information Operations. 9 October 1998.

Joint Pub 3-61. Doctrine for Public Affairs in Joint Operations.

Joint Pub 1-02. DOD Dictionary of Military and Associated Terms. 24 March 1994.

DOD Directive 5122.5 Public Affairs Program. 12 February 1993.

DOD Directive 5120.20-R. Management and Operation of Armed Forces Radio and Television Service. 9 November 1998.

FM 3-61 (46-1). Public Affairs Operations. 10 March 1997.

FM 3-100.71 (71-100). Division Operations. 28 August 1996.

FM 3-07.29 (90-29). Noncombatant Evacuation Operations. 17 October 1994.

FM 3-0 (100-5). Operations. 14 June 1993.

FM 3-13 (100-6). Information Operations. 27 August 1996.

FM 3-100.1 (100-15). Corps Operations. 29 October 1996.

FM 3-07.7 (100-19). Domestic Support Operations. 1 July 1993.

FM 3-07.6 (100-23). Peace Operations. 30 December 1994.

TRADOC PAM 525-5. Force XXI Operations. 1 August 1994.

AR 200-1. Environmental Protection and Enhancement. 23 April 1990.

AR 360-5. Public Information. 31 May 1989.

AR 360-61. Community Relations. 15 January 1987.

AR 360-81. Command Information Program. 20 October 1989.

Public Affairs Guidance on National Guard Bureau Environmental Programs, National Guard Bureau Office of Public Affairs, 1994.

Commander's Guide to Environmental Management, U.S. Army Environmental Center, 1995.

Index

Made in the USA
Monee, IL
04 November 2026

67382868R00095